SELECTED WORKS IN BIOINFORMATICS

Edited by **Xuhua Xia**

Selected Works in Bioinformatics
Edited by Xuhua Xia

Published by InTech
Janeza Trdine 9, 51000 Rijeka, Croatia

Publishing Process Manager Silvia Vlase

Technical Editor Teodora Smiljanic

Cover Designer InTech Design Team

Additional hard copies can be obtained from orders@intechopen.com

Selected Works in Bioinformatics, Edited by Xuhua Xia
p. cm.
ISBN 978-953-307-281-4

For EU product safety concerns:
IN TECH d.o.o., Prolaz Marije Krucifikse Kozulić 3, 51000 Rijeka, Croatia,
info@intechopen.com or visit our website at intechopen.com.

Meet the editor

Dr. Xuhua Xia was born in Jiangxi, China, obtained his PhD in population biology from University of Western Ontario, and worked as a postdoctoral fellow/visiting scientist at University of Helsinki, University of Toronto, University of Washington, Louisiana State University and University of Chicago, carrying out research in molecular biology and evolution, molecular phylogenetics, and population genetics. He became an assistant professor at University of Hong Kong in 1996, and served in 2001 as a senior scientist and head of Bioinformatics Laboratory in the newly established HKU-Pasteur Research Centre. He came back to Canada in 2002 and has been full professor at University of Ottawa since 2009. Dr. Xia published two books in bioinformatics and molecular evolution, and is the author of the software package DAMBE that is widely used in molecular biology and evolution. He currently teaches courses in bioinformatics, molecular evolution and biostatistics.

Contents

Preface

Bioinformatics has come a long way from its humble beginning of heuristic search engines such as FASTA and BLAST. The power of these heuristic search engines has resulted in rapid proliferation of specialized databases such as Pfam, InterPro, Prosite, UniProt, COG, CDD, from which a search will produce not only one or more matched sequences, but also a variety of functional annotations related to the query sequence. The database proliferation is an ongoing process, with some protein families (e.g., the G protein-coupled receptors) having their own database and website. These databases constitute essential tools for many biologists in their daily research activities.

The first chapter in this book, by Debajyoti Ghosh and Swati Gupta-Bhattacharya, brings us to two specialized databases of protein allergens as well as bioinformatic tools for the characterization, identification, and prediction of such protein allergens. Perhaps not many people still remember the transgenic soy that has been genetically engineered to express ground-nut 2S albumin which turned out to be a strong allergen for ground-nut allergic people. Such problems could have been avoided if searchable allergen databases were available. While only a very small number of proteins are known to be potent allergens - the allergome database currently lists less than 2000 protein allergens - it is quite possible that a protein that does not cause allergic response by itself, but it may well increase the allergic potency of other proteins. An allergic response may be caused by two or more proteins that are individually not allergenic. Allergen databases serve as essential infrastructure not only for future allergen research, but also for transgenic experiments in general.

Publicly available databases also facilitate the identification of disease genes, especially those involving multi-gene disorders. For example, conventional genetic linkage analysis based on human pedigrees or from breeding experiments using animal disease models allow scientists to map the disease trait to one or a few genomic regions. Sequencing these regions and searching them against databases will yield information on the differences between these regions and those in the database, whether these differences are in conserved or variable regions, in coding sequences or in regulatory sequences, whether the matched sequences/genes are already annotated to be related to the complex disease trait, what biological functions are performed by the gene products encoded in these regions, etc. The chapter by Christopher Vidal and

Angela Xuereb Anastasi provides a real example of such an application involving osteoporosis.

While database creation, maintenance and development continue to consume the time and effort of many bioinformaticians, most bioinformatic research projects are now focused on the identification of genetic switches, and how these switches are recognized by cellular machinery enabling the genetic program encoded in the genome to unfold during ontogeny. What switches are responsible for turning on the cellular machinery for genome replication and for its quality control? What switches are flipped on to initiate new protein synthesis in response to environmental challenges? What recognition system is packaged in the egg to turn on the first set of genetic switches in a newly formed zygote? Will we eventually be able to incorporate these genetic switches and the associated recognition system in a digital cell as a platform for drug testing, such as exploring the side effects of new biopharmaceuticals? What genetic switches often differ among people? How can the understanding of these differences in genetic switches among people guide us towards personalized medicine? It is not an exaggeration to state that modern molecular biology, which is the driving force for bioinformatics, is little more than the elucidation of genetic switches and their interactions with their dynamic recognition system in specifying the developmental process. Most chapters in this book deal with the recent development in characterizing and predicting the genetic switches and how they interact with other cellular components.

The most fundamental genetic switches are the regulatory sequence motifs in both nucleotide and amino acid sequences. The most well known types of nucleotide motifs include the origin of genome replication, the transcription factor binding site, the 5′ and 3′ splice sites and branchpoint site in eukaryotic introns, the Shine-Dalgarno sequence for prokaryotic translation initiation, the Kozak consensus for eukaryotic translation initiation, internal ribosomal entry site, the poly-A site, the numerous hormone response elements and many more. Well known peptide motifs include those that facilitate posttranslational modifications such as phosphorylation, acetylation/deacetylation, proteolytic cleavage, N-terminal processing, leader sequences for targeting various cellular and extracellular locations and so on.

Given a set of aligned sequence segments that are experimentally verified to be recognized by cellular machinery to accomplish a particular function, e.g., a set of transcription factor binding sites, how can we characterize these segments and use the characterization to scan the genome for new transcription factor binding sites? Phu Vuong and Rajeev Misra reviewed computational methods for identifying transcription factor binding sites in prokaryotic genomes. While their illustration of the method is somewhat primitive and does not really touch upon the modern computational approaches (in particular those incorporating site-dependence) and the associated statistical tests, the compilation of the relevant tools from a user perspective should be quite useful to readers who need a simple introduction to bridge them to more advanced bioinformatic tools and algorithms.

The chapter by Sergio Lobos and colleagues represents a more systematic application of motif-discovery methods to identify functional motifs in promoters of genes that participate in lignin degradation in *Phanerochaete chrysosporium*. The evolutionary perspective and the phylogenetic relationship among *lip* (lignin peroxidase) genes offer the reader a glimpse of how genetic switches may change during evolution leading to functional divergence.

While many bioinformatic studies focus on identifying shared sequence features in motifs and how these shared features are recognized by protein or RNA domains to accomplish a function, some sequences may share the same function but exhibit no similarity in either sequence or structure. There are two known types of functional sequences that belong to this category. The first is the eukaryotic ribosomal entry site (IRES). Among experimentally validated eukaryotic IRESs, no shared sequence or structural similarity has been found. Recent studies suggest that the shared feature is the lack of secondary structure, which explains their shared lack of sequence or structural similarities. The second type is exemplified by the intrinsically unfolded protein or region (IUP/IUR) which, according to the "dog-on-a-leash" model, may function by allowing a biochemically active domain (the dog) to stretch out to reach its target/substrate. The chapter by Cesira de Chiara and Annalisa Pastore provides an excellent introduction to IUP/IUR.

A biological concept closely related to regulatory motifs is the regulon, which is a set of genes sharing the same regulatory motif and co-regulated by the same cellular or environmental signal. For example, when oestrogen molecules diffuse into the cell and into the nucleus, they will bind to the oestrogen receptors which in turn will control the expression of genes sharing the oestrogen response element. The genes that share the oestrogen response element and co-regulated by the oestrogen through the oestrogen receptors constitute a regulon (the oestrogen regulon). The classical regulon is the *lac* operon in *Escherichia coli* that responds to the environmental availability of glucose and lactose.

How can we identify a regulon in response to an environmental change? In the case of the oestrogen regulon, we may only need to characterize the oestrogen response elements, and then scan the genome for other genes harbouring such elements to obtain a putative oestrogen regulon, followed by experimental validation. But what can we do if we do not know anything about the oestrogen response element? The common approach is to subject the experimental animals to oestrogen treatment, check which genes are up-regulated or down-regulated relative to the control by microarray, SAGE or deep sequencing method, and verify whether some co-expressed genes are in fact co-regulated, e.g., whether they share the same regulatory motif. Such a seemingly straightforward task is made difficult because of the immense noise inherent in the experimental result from large-scale transcriptomic studies. In addition to the genes in the oestrogen regulon that may increase or decrease their expression in response to the oestrogen treatment, many other genes may also increase or decrease their expression, either stochastically or through indirect/downstream regulation. How can we peel off the shell of noise and reach the central core of truth?

The chapter by Leon Bobrowski and Tomasz Łukaszuk, advocating the relaxed linear separability (RLS) approach to feature (gene) subset selection, represents one of the exciting new mathematical and computational tools that can help us to progressively narrow down to a minimal set of genes that lead to the best separation between the experiment and the control. Such a set of genes should greatly improve the efficiency of identifying shared regulatory motifs and co-regulated genes.

A conceptually closely related but computationally different method is presented by Carlos Roberto Arias, Hsiang-Yuan Yeh and Von-Wun Soo, in the context of disease gene prioritization. Given a set of genes, how would we know which one is the most related to the disease? Suppose we have two set of people, one with a particular disease and the other not. Also suppose that biologists have done a variety of phenotypic (e.g. gene expression) and genetic analysis (e.g., linkage analysis from pedigrees) related to the disease. Now given all these information (the training information) and a set of candidate genes (which could be large) that might be causative agents of the disease, how can we prioritize the genes to facilitate our identification of the genes responsible for the disease? The chapter by Arias et al. covers several approaches for such a complex task by using network methods.

What research should bioinformaticians be busy with in the future? The discovery of DNA and base complementarity paved the way for understanding and predicting interactions between nucleotide sequences – they interact mainly through complementary base pairing. The characterization of nucleotide regulatory motifs has led to fantastic progresses in predicting interactions between nucleotide sequences and proteins, such as that between splice sites and spliceosomes, or between restriction sites and restriction enzymes. However, understanding the interaction among proteins, the ultimate workhorses in the cell, still lags behind.

To be fair, biomedical scientists in general, and biochemists in particular, have characterized a large number of protein interactions, e.g., the interaction between antibody and antigen, between surface proteins of viruses (such as HIV-1) and the receptor proteins on the host cell membrane, between certain proteases and their cleavage sites, between various kinases and their target proteins, between acetylases/deacetylases and histones, etc. However, these interactions are not sufficient to understand how living cells work. We need large-scale methods to systematically characterize and predict all protein interactions in a cell and to understand how proteins come together to form partners through their surface features.

The last two chapters of the book are exclusively on proteins. A variety of technologies for large-scale characterization of protein interactions have been developed in recent years. The chapter by Bernd Sokolowski and Sandra Orchard reviews the technology and illustrate how the results can be used in conjunction with relevant databases to understand protein interactions and to construct protein interaction networks. Some readers might feel concerned about an overemphasis on physical interactions in recent years because many proteins cooperate to accomplish a task not by physical interactions. For example, the β-galactosidase (LacZ) and β-galactoside permease (LacY) encoded in the *E. coli lac* operon cooperate to accomplish the task of metabolizing lactose, but they do not physically interact. Similarly, not all Krebs cycle enzymes are found physically interacting. We need parallel development of biotechnological and bioinformatic methods for both functionally interacting networks and physically interaction networks. Some of the databases reviewed in the chapter are relevant for constructing functionally interacting networks.

With a large number of protein physical interactions and interaction networks documented in recent years, biologists have found themselves in an awkward position of knowing little about how proteins interact physically and specifically with their partners through their surface features. How much of the cellular protein interaction is due to surface geometry, electrostatic interaction or simply due to local crowding that forces together otherwise non-interacting proteins? It would be really nice to know how interacting protein partners find each other, adapt their surface features to each other and achieve the structural docking necessary for their function. The time is ripe for such research endeavour. Recent increase in the number of protein structures has provided an excellent platform for validating computational algorithms for structural modelling. In particular, structures of protein-protein complexes and protein-RNA complexes have shed light on the induced structural changes of proteins and RNA in response to the surface features of their interacting partners. The chapter by Sebastian Schneider and Martin Zacharias offers an advanced review of the current computational algorithms used in flexible protein-protein docking.

Dr. Xuhua Xia
Biology Department
University of Ottawa
Canada

1

Allergen Bioinformatics: Recent Trends and Developments

Debajyoti Ghosh[1] and Swati Gupta-Bhattacharya[2]
[1]*Division of Allergy, Immunology and Rheumatology, Department of Internal Medicine*
University of Cincinnati College of Medicine, Ohio
[2]*Division of Plant Biology Bose Institute Kolkata*
[1]*United States of America*
[2]*India*

1. Introduction

Allergy is a major cause of morbidity worldwide. Allergic reactions result from maladaptive immune responses in predisposed subjects, to otherwise harmless molecules. These allergenic molecules, usually proteins/glycoproteins, can not only elicit specific immunoglobulin E (IgE) in susceptible subjects, but also crosslink effector cell-bound IgE molecules Leading to the release of mediators (e.g. Histamine) and causation of symptoms. From clinical and molecular biological data available in several publicly accessible databases, it is now evident that among hundreds and thousands of proteins that exist in nature, only a few can cause allergy. For example, in more than 500,000 entries (71345 documented at the protein level; Nov, 2010) in swissprot/uniprot database (http://www.uniprot.org), only 686 proteins have been listed in the IUIS allergen nomenclature database (www.allergen.org) as documented allergens. Although about 1500 allergens (including iso-allergens) have been listed in the Allergome database (www.allergome.org), it has been shown that they are distributed into a very limited number of protein families. However, critical feature(s) that makes proteins allergenic is not fully understood. In the present article, we'll discuss recent applications of bioinformatic tools that shaped our current understanding about allergenicity of proteins.

2. Allergen bioinformatics - a need of the hour

Experiments on genetic engineering during the last few decades have led to the production of numerous genetically modified (GM) organisms. So, proteins introduced into GM organisms through genetic engineering must be evaluated for their potential to cause allergic diseases. As a classical example, transgenic soy, that has been genetically engineered to express ground-nut 2S albumin, was found to elicit hypersensitivity reactions in ground-nut allergic people (Nordlee et al., 1996). In 2001, the FAO/WHO suggested a procedure for performing FASTA or BLAST (Basic Local Alignment Search Tool) searches, and a threshold of greater than 35% identity in 80 or greater amino acids to identify potential allergenic cross-reactivity of transgene encoded proteins in genetically enhanced crops (Silvanovich et

al., 2009). Given that this will not exclude all probabilities of a protein to be allergenic (and cross-reactive to known allergens), the codex guidance recognized that the assessment will evolve based on new scientific knowledge (Goodman, 2008).
Bioinformatic tools are key components of the 2009 Codex Alimentarius for an overall assessment of the allergenic potential of novel proteins. Bioinformatic search comparisons between novel protein sequences, as well as potential novel fusion sequences derived from the genome and transgene or from any known allergen(s) are required by all regulatory agencies that assess the safety of genetically modified (GM) products(Ladics et al., 2011).
Allergens were usually seen as an array of proteins with no apparent similarity in structure and function. They come from diverse sources: Plants, animals or fungi and may take different modes of exposure: inhalation, ingestion, sting or contact. They are, like their non-allergenic counterparts, structurally heterogeneous. For example, the major cat allergen Fel d 1, is an alpha-helical tropomyosin, while a major dust mite allergen Der p 2 consists predominantly of beta sheets and the major birch pollen allergen Bet v 1 contains both of these structural elements. Allergen sequences are extensively studied to find out any possible structural element or function associated with allergenicity. However, no such allergen-specific structural / functional element could be identified. High sequence identity between homologous protein allergens may result in common surface patches that may confer cross-reactivity among them. Aalberse pointed out that proteins sharing less than 50% sequence identity are rarely cross-reactive(Aalberse, 2000). In contrast, proteins that share at least 70% identity often show cross-reactivity. Many IgE-binding epitopes have been identified as sequential epitopes, although for many this does not represent the full epitope. Linear epitopes are usually part(s) of conformational epitope(s) responsible for a significant portion of IgE binding. While IgE-binding peptides can consist only of five amino acids (Banerjee et al., 1999), the majority of characterized IgE-linear epitopes are eight amino acids or longer (Chatchatee et al., 2001; Shin et al., 1998). Astwood et al. recommended sequence comparisons to a database of known IgE-binding epitopes. Finally, Ivanciuc and colleagues have recently utilized mixed sequence and structure-based methods to predict IgE-binding sites. This is based on comparison of local sequence and structure to identify common features associated to allergens (Ivanciuc et al., 2009b).

3. Allergen databases

Exponential growth of molecular and clinical data on allergens has created a huge demand for efficient storage, retrieval and analyses of available information. There are numerous allergen databases available on Internet. They are targeted to different aims ranging from easy accessibility of data to novel allergen prediction. A few examples have been provided in table-1.
The IUIS (International Union of Immunological Societies) allergen nomenclature subcommittee has created a unique, unambiguous nomenclature system for allergenic proteins. It maintains an allergen database (www.allergen.org) containing an expandable list of WHO/IUIS -recognized allergen molecules arranged according to Linnean system of classification (Kingdoms: Plantae, Fungi and Animalia and subdivided into lower orders)(Chapman et al., 2007) of the source organism. This database is a precise and convenient source for researchers, since it contains the biochemical name and molecular weight of the allergens and isoallergens (multiple molecular forms of the same allergen showing ≥67% sequence identity). It is searchable by allergen name, source and taxonomic

group. For example, a search using the key word 'Bet v 1' shows about 36 variants (isoallergens) of this allergen, each with genbank, uniprot accession numbers and, if available, with PDB IDs. Each uniprot ID is linked to the original entry in uniprot database. Moreover, once the uniprot IDs are obtained, their sequences can be retrieved in batches using uniprot's 'retrieve' tab.

Allergome (Mari et al., 2006) is a vast repository of data related to all allergen molecules. It contains data about a larger number of allergens than actually recognized by IUIS/WHO. It also contains links to other databases (eg Uniprot, PDB) and computational resources with additional extensive links to literature. The Allfam database is a useful resource for grouping of allergens into protein families. It utilizes the allergen information from 'Allergome' database and protein family information from pfam database. It can be sorted by source (plants/animals/bacteria/fungi) and route of exposure (inhalation/ingestion/contact/sting etc) or can be searched for specific protein families. Allergen entries are linked to corresponding records in the Allergome database. In addition, each allergen family is linked to a family fact sheet containing descriptions of the biochemical properties and the allergological significance of the family members.

Name (URL)	Purpose
IUIS (http://www.allergen.org/)	Database targated towards systematic nomenclature of allergenic proteins
Allergome (http://www.allergome.org/)	Vast source of information and references about allergen molecules
Allfam (http://www.meduniwien.ac.at/allergens/allfam/)	Database for allergen classification
Allergen Database for Food Safety (ADFS) (http://allergen.nihs.go.jp/ADFS/)	Database with computational allergenicity prediction tool
The Allergen Database (http://allergen.csl.gov.uk//index.htm)	A basic database for allergen structures
Allermatch (http://www.allermatch.org/)	Allergenicity prediction from sequence
AllerTool http://research.i2r.a-star.edu.sg/AllerTool/	Webserver for predicting allergenicity and allergenic cross-reactivity
AlgPred (http://www.imtech.res.in/raghava/algpred/)	*In silico* prediction of allergenicity
WebAllergen (http://weballergen.bii.a-star.edu.sg/)	To predict potential allergenicity of a protein from its sequence
SDAP (http://fermi.utmb.edu/SDAP/)	Database of allergen structure with various resources, links and computational tools

Table 1. A few databases of allergenic proteins and web-servers to predict protentialallergenicity from amino-acid sequence.

Although the above-mentioned databases are very useful resources, they do not contain any computational tool to predict allergenicity from amino acid sequences of proteins. However, there are several other databases that can efficiently deal with this aspect. ADFS (Allergen Database for Food Safety) is developed and maintained by Japan's National Institute of Health Sciences. It is a good resource of available information about known allergens (uniprot protein ID, PDB accession number, epitope sequence, presence of carbohydrate, pfam - and interpro domain IDs etc.). Moreover, this website has computational tools to predict allergenicity. Other websites dedicated to allergenicity prediction are Allermatch, AllerTool and Algpred etc. Detailed discussion on these servers is beyond the scope of the present article.
The database which is dedicated to the structural biology of the allergic proteins is SDAP (Structural Database of Allergenic Proteins) hosted by the University of Texas Medical Branch. It integrates a database of allergenic proteins with various computational tools for prediction of allergenicity and epitope sequences on protein allergens.
Analyses of data available in different publicly accessible database have shaped our current understanding about allergens, as discussed in the subsequent sections of this article.

3.1 Allergens seen as proteins without bacterial homolog

Among numerous proteins sequenced till date, only about a thousand has been classified as allergens, although no common structural or biochemical function could be assigned to all allergens. To address this problem, Emanuelson and Spangfort (2007) used 30 randomly selected allergen sequences to search the non-redundant Expasy/SIB and UniProt/TrEMBL databases (subsection Bacteria+Archea) using BLAST (Basic Local Alignment Search Tool) program. For each allergen, an appropriate species-specific non-allergenic control homolog was included. It has been found that 25 out of 30 allergens do not have any bacterial homologues; two other allergens have only a few, while all the non-allergenic controls retrieved numerous bacterial homologues. Moreover, major allergens like Bet v 1, also lack human homolog. The authors, thus, interpreted that the allergens are usually foreign proteins that lack bacterial homologues (Emanuelsson and Spangfort, 2007).

3.2 Allergenic proteins can be organized into families

The first definite interpretation that allergens can be grouped came from arranging allergens into pfam protein families. Pfam classifies proteins into families on the presence of specific domains (pfam domains) identified through multiple sequence alignments and Hidden Markov Models. Pfam 25.0 (latest version; March 2011) contains over 100, 000 protein sequences classified into 12,275 families (Finn et al., 2010). The allergen database that contains pfam domain information is 'AllFam' (http://www.meduniwien.ac.at /allergens/allfam), where allergen sequences are classified into protein families using the Pfam database, and its associated database, SwissPfam. AllFam includes all allergens that can be assigned to at least one Pfam family. But many allergens are multi-domain proteins. The domains of these proteins are merged into a single AllFam family, if the Pfam domains of this allergen occur only in combination with a single other Pfam domain. Figure-1 shows the distribution of allergenic proteins in different Allfam families. The major allergen families (containing 10 or more allergens) with corresponding Pfam domains are shown in Table-2.

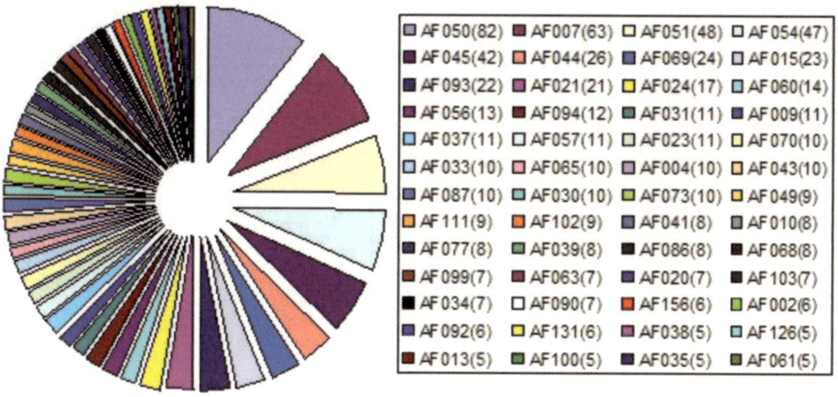

Fig. 1. Pi chart showing the distribution of allergic proteins in different AllFam families. Numbers of constituent allergens have been indicated within brackets.

Allfam ID	Name	Pfam ID	Allergens	Examples
AF050	Prolamin superfamily	PF00234	82	Amb a 6 (Ragweed) Ana o 3 (cashewnut)
AF007	EF hand domain	PF00036 PF01023	63	Aln g 4 (Alder) Bet v 3 (birch)
AF051	Profilin	PF00235	48	Bet v 2 (birch) Ana c 1 (pinapple)
AF054	Tropomyosin	PF01357	47	Der p 10 I(mite) Hom a 1 (lobster)
AF045	Cupin superfamily	PF00190 PF04702	42	Ara h 1 (peanut) Gly m 5 (soyabean)
AF044	CRISP/PR-1/venom group 5 allergen family	PF00188	26	Pol d 5 (wasp venom), Art v 2 (mungwort)
AF069	Bet v 1-related protein	PF00407	24	Bet v 1 (birch) Api g 1 (celery)
AF015	Lipocalin	PF00061 PF08212	23	Can f 1 (dog) Bos d 2 (domestic cattle)
AF093	Expansin, C-terminal domain	PF01357	22	Phl p 1 (timothy grass) Tri a 1 (wheat)
AF021	Subtilisin-like serine protease	PF00082 PF02225 PF05922	21	Asp f 13 (fungal) Pen c 1 (fungal)
AF024	Trypsin-like serine protease	PF00089 PF02983 PF09396	17	Der f 3 (mite) Blo t 3 (mite)
AF060	Thaumatin-like protein	PF00314	14	Mal d 2 (apple) Pru av 2 (Cherry)
AF056	Serum albumin	PF00273	13	Can f 3 (dog) Fel d 2 (cat)

AF094	Expansin, N-terminal domain	PF03330	12	Phl p 1 (timothy grass) Ory s 1 (rice)
AF031	Enolase	PF00113 PF3952	11	Alt a 6 (fungal) Cha h 6 (fungal)
AF009	Globin	PF00042	11	Chi t 1 (midge) Chi t 2 (midge)
AF043	Hevein-like domain	PF00187	11	Hev b 6 (rubber latex) Mus a 2 (banana)
AF073	Pectate lyase	PF00544	11	Amb a 1 (ragweed) Cry j 1 (Japanese cedar)
AF057	Polygalacturonase	PF00295	11	Cry j 2 (Japanese Cedar) Jun a 2 (mountain Cedar)
AF037	Lipase	PF00151 PF01477	11	Pol a 1 (wasp) Sol i 1 (ant)
AF070	60S acidic ribosomal protein	PF00428	10	Alt a 5 (fungal) Cla h 6 (fungal)
AF033	Alpha amylase	PF00128 PF02806 PF07821 PF09154 PF09260	10	Blo t 4 (mite) Der p 4 (mite)
AF065	Alpha/beta casein	PF00363	10	Bos d 8 alphaS1 (domestic cattle) Ovi a casein alphaS1 (sheep)
AF004	Eukaryotic aspartyl protease	PF00026 PF07966	10	Asp f 10 (fungal) Bla g 2 (cockroach)
AF030	Papain-like serin protease	PF00112 PF08246	10	Der p 1 (mite) Blo t 1 (mite)
AF023	Thioredoxin	PF00085	10	Alt a 4 (fungal) Fus c 2 (fungal)
AF073	Pectate lyase	PF00544	10	Amb a 1 (short ragweed) Jun a 1 (mountain ceder)

Table 2. Major allergen families (AllFam familes) that contain 10 or more allergens are shown with correspondent pfam domains and examples. Number of allergenic members / allergen family has also been shown.

AllFam takes the allergen information from "Allergome", the comprehensive allergen database. In the latest version of Allfam (May, 2011), 950 allergens have been arranged into 150 allergen families (AllFam families). It has been found that the allergens are distributed in a really skewed manner with about 30% members belonging to only 5 families (Prolamin, Profilins, EF hands, tropomyosin and cupins) and showing few restricted biological functions such as hydrolysis, storage or binding to cytoskeleton [6]. Moreover, allergens contain about 245 pfam domains in total, which is only about 2.0% of all domains identified to date.

AllFam gave us an opportunity to retrieve and sort allergen data according to source (plant/animal/fungi/bacteria), route of exposure (inhalation/ingestion/contact etc) and

Pfam/AllFam family identities. This analysis combined with the study of evolutionary relationship among the proteins has led to the following valuable insights:

i. Pollen allergens (Inhalant plant allergens) are restricted into few protein families(Radauer and Breiteneder, 2006). They populate only 29 out of more than 7000 protein families, with (a) Expansins (b) Profilins and (c) calcium-binding proteins (with EF-hand domains) consisting most of the pollen allergens followed by Bet v 1 related /pathogenesis-related proteins (PR10 family). Figure-2 shows the evolutionary relationship between several allergenic and non-allergenic members of (a) expansins and (b) profiling families. The evolutionary history was inferred using the Neighbor-Joining method(Saitou and Nei, 1987). The evolutionary distances were computed using the Poisson correction method(Zuckerkandl and Pauling, 1965) and the phylogenetic analyses were conducted in MEGA4(Tamura et al., 2007). Similar method has been followed in the subsequent sections of the present article. Allergens of the expansin family are clustered as highly identical proteins as shown in the figure. Allergenic plant profilins also constitute a conserved homologous group with high sequence identities (70-85%) among themselves, while showing low identities (30-40%) with non-allergenic profilins from other eukaryotes including human(Radauer and Breiteneder, 2006). About 10 of the 29 pollen allergen families are also present in plant-derived foods.

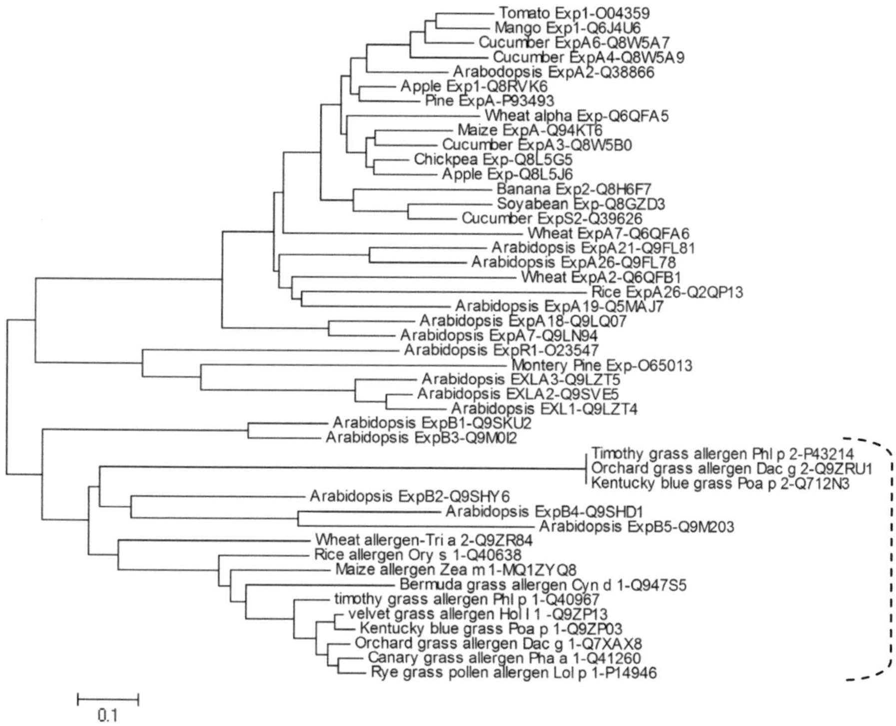

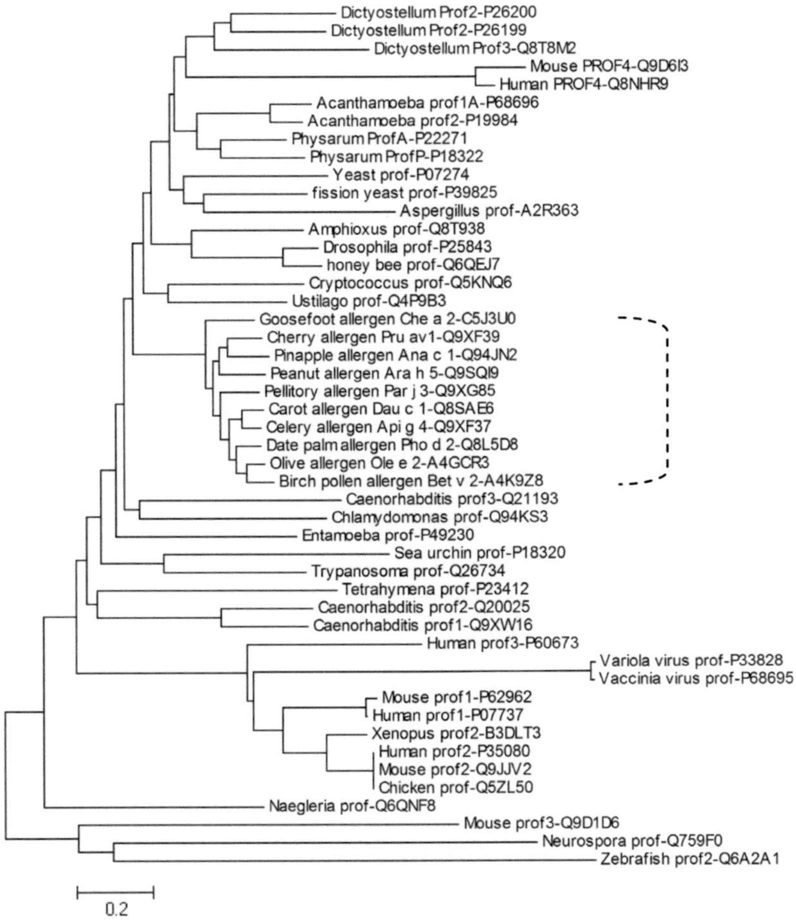

Fig. 2. Phylogenetic trees showing the relationships of two major pollen allergen families: (a) Expansins, (b) Profilins and their respective non-allergenic homologues. Pollen-related plant food allergens such as Ara h 5, Dau c 1 etc are also included. Uniprot accession numbers are shown. Positions of allergens are indicated by dotted lines.

ii. In case of major animal food protein families evolutionary distance from human homologue reflects their allergenicity (Jenkins et al., 2007). This has been demonstrated in major food allergen families like (a) parvalbumins, (b) casins and (c) tropomyosins.

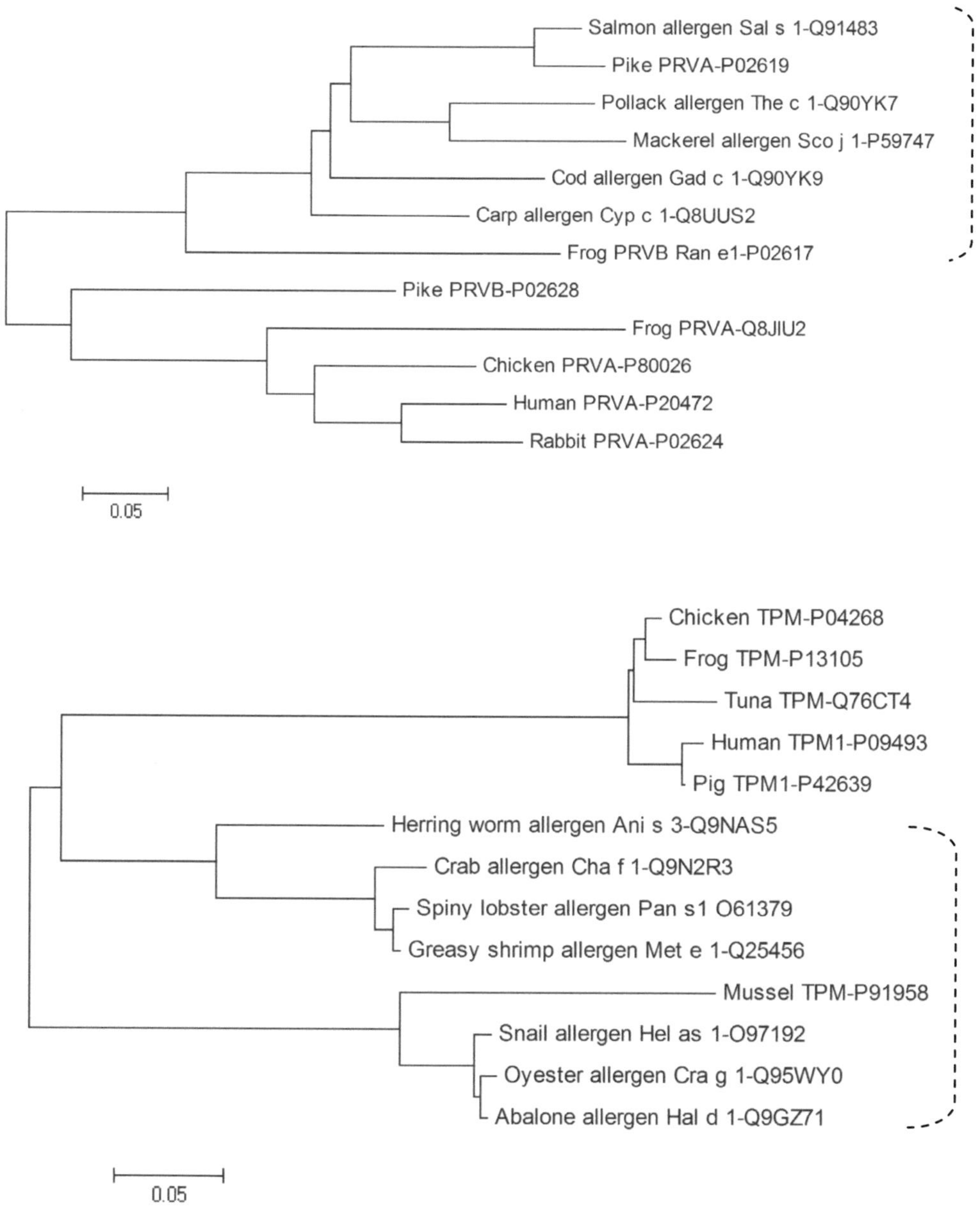

Fig. 3. Dendrogram showing evolutionary relationship among 12 different parvalbumins (a) and 13 different tropomyosins (b) from animals and human. Allergenic proteins and their non-allergenic homologues as well as the closest human homologues are chosen. The Uniprot accession numbers and positions of allergen clusters are indicated.

iii. Plant food allergens are clustered into only four major protein families

Fig. 4. Un-rooted neighbor-joining tree showing evolutionary relationship among the members of Bet v 1-related plant protein family (containing pfam domain PF00407). Uniprot accession numbers and position of the allergen cluster has been indicated.(Radauer and Breiteneder, 2007). They are (a) the Prolamin superfamily with PF00234 domain (b) the cupin superfamily with PF00190 and PF04702 domains (c) the Profilins with PR00235

domain and (d) the Bet v 1 -like proteins containing PF00407 domain. Prolamins are seed storage proteins containing about 82 characterized allergens, with 65 enlisted as ingestants. Figure-4 shows the evolutionary relationship among the Bet v 1-homologous protein family. Twenty-four proteins of this group are known as allergens present in pollen and plant-derived foods responsible for causing allergic sensitization in a large number of people.

3.3 Allergen-associated protein domains

The other allergen database that utilizes the Pfam protein family information is Motifmate (http://born.utmb.edu/motifmate/index.php) (Ivanciuc et al., 2009a). Motifmate assigns pfam domains to the allergens listed in the SDAP (Structural Database of Allergenic Proteins) database developed and maintained by the University of Texas (http://fermi.utmb.edu/SDAP)(Ivanciuc et al., 2003). The recent version of this database contains 679 Proteins (May, 2011). The authors pointed out that all the allergenic protein entries in SDAP could be associated with only 130 pfams (of total 9318 pfams) with only about 31 pfam protein families containing 4 or more number of allergens. [Figure-5]. This outcome supports the previous finding that the allergenic proteins are clustered in few pfam families.

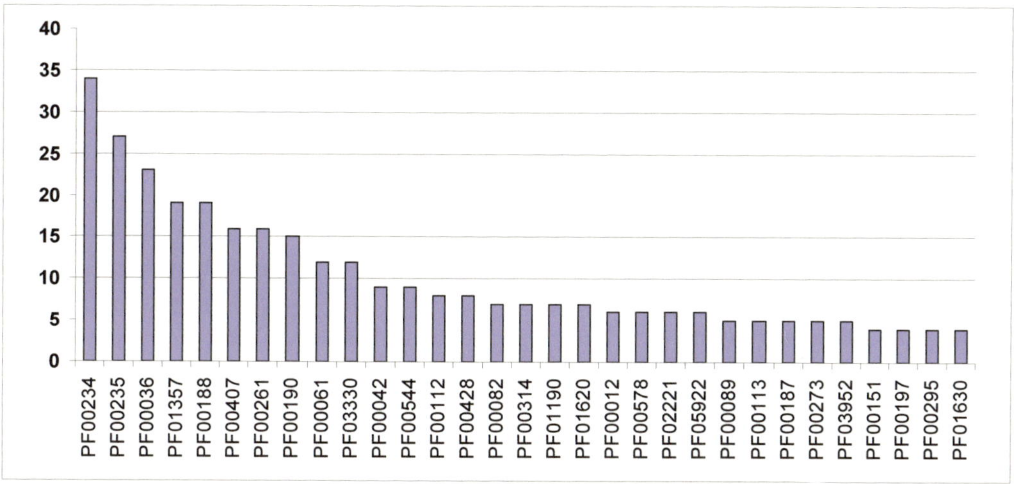

Fig. 5. Distribution of Pfam domains in allergenic proteins according to Motifmate database. lipid transfer proteins (PF0234) highest number of allergns.

4. Insights from structural bioinformatics

After the elucidation of X-ray crystal structure of the birch pollen allergen Bet v 1(Gajhede et al., 1996), structures of several allergens have been solved. Searching the protein databank with the keyword "allergens" returns 321 entries, with occasional presence of multiple entries for one single allergen. Although protein structure gives us valuable insight into their function, structures of several allergens are still not known. More importantly, some allergen families have members with known structures, while others may have very few / no member whose structure(s) have been deduced. Allergen structures are particularly useful to elucidate molecular features related to allergenicity, cross-reactivity and for designing hypoallergenic derivatives. For example, there are about five structures in protein data bank that correspond to Bet v 1, the major birch pollen allergen : the x-ray structure (1BV1.pdb), the NMR structure (1BTV.pdb), mutants (1B6F.pdb and 1QMR.pdb), complexed with IgG Fab (1FSK.pdb) and the hypoallergenic isoform Bet v 1d (3K78.pdb). On the contrary, several groups, such as the cupin family of seed storage protein allergens are under-represented.

Knowledge about allergen structures is important because it is the over-all structure, not the sequence, which determines the biochemical/immunological properties. Molecular modeling can help us in case the experimentally determined structure of the allergen is not available. Homology modeling, also known as comparative molecular modeling, can predict the 3D model of a given protein from its amino acid sequence using experimentally derived structure(s) (X-ray/NMR) of one or more related homologous protein(s) (called template). This technique is becoming increasingly popular because, if required template selection and alignment criteria are met, it is believed to be the most reliable modeling technique to date (Marti-Renom et al., 2000). It is becoming increasingly useful because although there are millions of proteins in nature, the number of structural folds they can assume is limited (Zhang, 1997) and the number of X-ray/NMR structure of proteins is exponentially increasing providing an increased chance of getting a suitable 'template'. Several authors have successfully utilized this technique of molecular modeling to predict allergen structures and to elucidate the structural basis of cross-reactivity between allergens.

Ara h 1 (vicilin) and Ara h 2 (2S albumin) are seed storage proteins of peanut (*Arachis hypogea*). They are recognized by serum IgE of >90% of peanut-allergic people, thus showing their importance as major peanut allergens (Shin et al., 1998; Stanley et al., 1997). Ara h 1 shows IgE-mediated cross-reactivity with other vicilin allergens such as Len c 1 (from lentil) and Pis s 1 (from sweet pea). Following sequence alignment using ClustalX, structural models of Ara h 1, Len c 1 and Pis s 1 were generated from experimentally derived structure of beta-conglycinin (RCSB protein data bank code: 1IPJ) using programs InsightII, Homology and Discover3 (Accelrys, USA). Electrostatic surfaces of these proteins were also generated using program GRASP (Nicholls et al., 1991). Mapping of linear epitope sequences revealed that nine out of 23 linear B-cell epitopes are located in the N-terminal region. They are unique to Ara h 1. But the remaining B-cell epitopes, situated in the C-terminal part, are well-exposed to the surface, share a high degree of homology and 3D conformation to Len c 1 and Pis s 1. They might be responsible for cross-reactivity among Ara h 1, Len c 1 and Pis s 1 food proteins. Similarly, Ara h 2 and other dietary allergenic 2S albumins Jug r 1 (walnut), Car i 1 (pecan nut), Ber e 1 (Brazil nut) were modeled using the atomic coordinates of homologous Ric c 1(castor bean 2S albumin). Mapping of known epitope sequences on the template and modeled structures revealed no structural homology between allergenic 2S albumins of peanut, walnut, pecan and brazil nut. This indicates that cross-reactivity between Ara h 1 and other 2S albumins, which is less likely, might not be mediated by protein epitopes, but CCDs (cross-reactive carbohydrate determinants). However, the c-terminal epitope region of Jug r 1 showed a clear structural homology with Car i 1 indicating the possibility of their cross-reactivity.

Another important insight was obtained from the homology modeling of allergenic cyclophilins(Roy et al., 2003). Groups of highly homologous cross-reactive allergens such as cyclophilins, profilins, MnSOD are known as pan-allergens. They often cross-react with their respective human homologues(Crameri et al., 1996) and such cross-reactivity might be responsible for severity and perpetuation of symptoms in the absence of exogenous allergen exposure(Fluckiger et al., 2002). Allergenic cyclophilins (peptidyl-prolyl cis-trans isomerase; PF00160) have been identified from several organisms such as: Periwinkle (pollen allergen Cat r 1), birch (pollen allergen Bet v 7), *Aspergillus fumigatus* (Asp f 11, Asp f 27), *Psilocybe cubensis, Malassazia furfur* (Mala s 6, formerly known as Mal f 6) and carrot. IgE-mediated cross-reactivity between Mala s 6, Asp f 11, yeast cyclophilin and human cyclophins has been demonstrated (Fluckiger et al., 2002). The structure of human cyclophilin, which shows high sequence identities to Asp f 11, Mala s 6 and yeast cyclophilin, was known from crystallography (PDB code: 2RMB). Thus, taking this as the template, the molecular models of three other cyclophilins were generated and compared with the human homologue to understand the structural basis of their cross-reactivity.

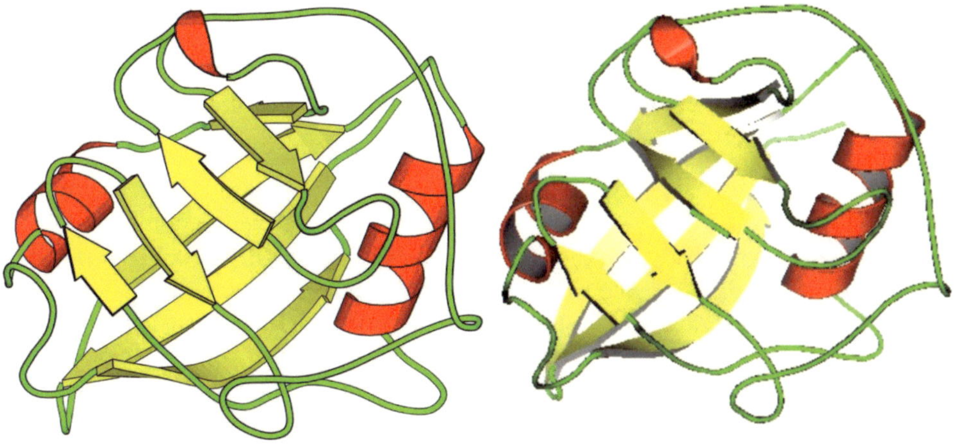

Fig. 6. Structure diagram of the allergenic cyclophilin Mala s 6 as predicted in 2003 from homology modeling (left figure) and as determined by x-ray crystallography (right figure) in 2006 (2CFE.pdb). Alpha helices have been shown in red, while beta sheets are in yellow and loops (smoothed) are in green.

Molecular modeling was done using program Modeller (Sali and Blundell, 1993). The structures were energy-minimized using program Discover with Consistent Valence Force Fields(Hagler et al., 1979) and their steriochemical qualities were checked using Procheck(Laskowski, 1993). Several empirical/semi-empirical programs were used to predict the antibody binding sites (B-cell epitopes) on these proteins and residue-wise solvent accessibility values of these predicted epitopes were calculated using program NACCESS(Hubber, 1992). The cyclosporine-binding site of these proteins were also identified by aligning the sequences (using ClustalW) and structures. This study revealed large conserved solvent-exposed patches on the surfaces of these proteins strongly suggesting their cross-reactivity. The x-ray crystal structure of Mala s 6 (PDB code: 2CFE), published three years later (Glaser et al., 2006), very much resembles its predicted model [Figure-6]. The domain-swapped structure of Asp f 11 dimer (PDB code: 2C3B), also published at the same time, showed similar structural fold. Asp f 11 dimer (resulting from increased protein concentration) seems to be enzymatically inactive, since the active sites of both its subunits are blocked due to dimerization. However the constituent monomers

retained the basic cyclophilin structure. More recently, a comprehensive 3D structural modeling of allergens, with no known structure, has been conducted followed by surface accessibility calculation and mapping of known IgE-binding epitope sequences. It has been found that Ala, Asn, Gly and Lysine have a high propensity to occur in the IgE-binding sites on the surface of allergenic proteins (Oezguen et al., 2008).
Finally, techniques of structural bioinformatics have also been applied to assess features critically required for allergenicity and cross-reactivity. This has been done by analyzing the predicted structure of protein T1, the naturally occurring non-allergenic member of the Bet v 1 allergen family (Ghosh and Gupta-Bhattacharya, 2008). Protein T1 shows considerable sequence similarity with the proteins of Bet v 1 allergen family, but it is neither allergenic, nor cross-reactive to the Bet v 1 group (Laffer et al., 2003). Comparative molecular modeling, solvent accessibility calculations and mapping of surface electrostatic potential showed substantial difference in antigenic surface that can be responsible for the loss of cross-reactivity. Solvent-accessible surface area and electrostatics calculations were done using program DSSP (dictionary of secondary structure of proteins) and program APBS (Adoptive Poissopn-Blotzman solver) respectively (Baker et al., 2001; Kabsch and Sander, 1983). Although, as suggested by ligand docking, it should be able to perform its biological function as a brassinosteroid carrier.

5. Conclusion

Allergy is a world-wide problem. Allergic symptoms are elicited following exposure to a structurally diverse group of proteins known as allergens. Understanding allergenicity at the molecular level has wide application in food safety and in treating allergic diseases. What makes a protein allergic is not yet understood. However, advanced tools of bioinformatics have been applied to address this problem. It has been found that allergens are usually foreign proteins with few/no bacterial homologue. They are clustered into few protein families (associated with a limited number of protein domains) opposing the idea that any protein can be allergenic. Methods have been developed to predict probable allergenicity from protein sequence, although more works need to be done for better and more precise prediction. The structural difference between IgG-binding and IgE-binding epitopes is still not very clear, but homology modelling in combination with residue-wise solvent-accessibility of monomers and biological assemblies of allergens certainly gives valuable information about antigenic determinants on protein allergens.

6. References

Aalberse R. C. (2000) Structural biology of allergens. *J Allergy Clin Immunol* 106, 228-38.

Baker N. A., Sept D., Joseph S., Holst M. J. and McCammon J. A. (2001) Electrostatics of nanosystems: application to microtubules and the ribosome. *Proc Natl Acad Sci U S A* 98, 10037-41.

Banerjee B., Greenberger P. A., Fink J. N. and Kurup V. P. (1999) Conformational and linear B-cell epitopes of Asp f 2, a major allergen of Aspergillus fumigatus, bind differently to immunoglobulin E antibody in the sera of allergic bronchopulmonary aspergillosis patients. *Infect Immun* 67, 2284-91.

Chapman M. D., Pomes A., Breiteneder H. and Ferreira F. (2007) Nomenclature and structural biology of allergens. *J Allergy Clin Immunol* 119, 414-20.

Chatchatee P., Jarvinen K. M., Bardina L., Beyer K. and Sampson H. A. (2001) Identification of IgE- and IgG-binding epitopes on alpha(s1)-casein: differences in patients with persistent and transient cow's milk allergy. *J Allergy Clin Immunol* 107, 379-83.

Crameri R., Faith A., Hemmann S., Jaussi R., Ismail C., Menz G. and Blaser K. (1996) Humoral and cell-mediated autoimmunity in allergy to Aspergillus fumigatus. *J Exp Med* 184, 265-70.

Emanuelsson C. and Spangfort M. D. (2007) Allergens as eukaryotic proteins lacking bacterial homologues. *Mol Immunol* 44, 3256-60.

Finn R. D., Mistry J., Tate J., Coggill P., Heger A., Pollington J. E., Gavin O. L., Gunasekaran P., Ceric G., Forslund K., Holm L., Sonnhammer E. L., Eddy S. R. and Bateman A. (2010) The Pfam protein families database. *Nucleic Acids Res* 38, D211-22.

Fluckiger S., Fijten H., Whitley P., Blaser K. and Crameri R. (2002) Cyclophilins, a new family of cross-reactive allergens. *Eur J Immunol* 32, 10-7.

Gajhede M., Osmark P., Poulsen F. M., Ipsen H., Larsen J. N., Joost van Neerven R. J., Schou C., Lowenstein H. and Spangfort M. D. (1996) X-ray and NMR structure of Bet v 1, the origin of birch pollen allergy. *Nat Struct Biol* 3, 1040-5.

Ghosh D. and Gupta-Bhattacharya S. (2008) Structural insight into protein T1, the non-allergenic member of the Bet v 1 allergen family-An in silico analysis. *Mol Immunol* 45, 456-62.

Glaser A. G., Limacher A., Fluckiger S., Scheynius A., Scapozza L. and Crameri R. (2006) Analysis of the cross-reactivity and of the 1.5 A crystal structure of the Malassezia sympodialis Mala s 6 allergen, a member of the cyclophilin pan-allergen family. *Biochem J* 396, 41-9.

Goodman R. E. (2008) Performing IgE serum testing due to bioinformatics matches in the allergenicity assessment of GM crops. *Food Chem Toxicol* 46 Suppl 10, S24-34.

Hagler A. T., Lifson S. and Dauber P. (1979) Consistant force field studies of intermolecular forces in hydrogen-bonded crystals.2. A benchmark for the objective comparison of alternative force fields. *J. Amer. Chem. Soc.* 101, 5122-5130.

Hubber S. (1992) ACCESS: A program for calculating Accessibilities. *Dept. of Biochemistry and Molecular Biology, Univ Col of London.*

Ivanciuc O., Garcia T., Torres M., Schein C. H. and Braun W. (2009a) Characteristic motifs for families of allergenic proteins. *Mol Immunol* 46, 559-68.

Ivanciuc O., Schein C. H. and Braun W. (2003) SDAP: database and computational tools for allergenic proteins. *Nucleic Acids Res* 31, 359-62.

Ivanciuc O., Schein C. H., Garcia T., Oezguen N., Negi S. S. and Braun W. (2009b) Structural analysis of linear and conformational epitopes of allergens. *Regul Toxicol Pharmacol* 54, S11-9.

Jenkins J. A., Breiteneder H. and Mills E. N. (2007) Evolutionary distance from human homologs reflects allergenicity of animal food proteins. *J Allergy Clin Immunol* 120, 1399-405.

Kabsch W. and Sander C. (1983) Dictionary of protein secondary structure: pattern recognition of hydrogen-bonded and geometrical features. *Biopolymers* 22, 2577-637.

Ladics G. S., Cressman R. F., Herouet-Guicheney C., Herman R. A., Privalle L., Song P., Ward J. M. and McClain S. (2011) Bioinformatics and the allergy assessment of agricultural biotechnology products: Industry practices and recommendations. *Regul Toxicol Pharmacol.* 60, 46-53.

Laffer S., Hamdi S., Lupinek C., Sperr W. R., Valent P., Verdino P., Keller W., Grote M., Hoffmann-Sommergruber K., Scheiner O., Kraft D., Rideau M. and Valenta R.

(2003) Molecular characterization of recombinant T1, a non-allergenic periwinkle (Catharanthus roseus) protein, with sequence similarity to the Bet v 1 plant allergen family. *Biochem J* 373, 261-9.

Laskowski R. A. (1993) PROCHECK: A program to check the stereochemistry of protein structure. *J. Appl. Cryst.* 26, 283-291.

Mari A., Scala E., Palazzo P., Ridolfi S., Zennaro D. and Carabella G. (2006) Bioinformatics applied to allergy: allergen databases, from collecting sequence information to data integration. The Allergome platform as a model. *Cell Immunol* 244, 97-100.

Marti-Renom M. A., Stuart A. C., Fiser A., Sanchez R., Melo F. and Sali A. (2000) Comparative protein structure modeling of genes and genomes. *Annu Rev Biophys Biomol Struct* 29, 291-325.

Nicholls A., Sharp K. A. and Honig B. (1991) Protein folding and association: insights from the interfacial and thermodynamic properties of hydrocarbons. *Proteins* 11, 281-96.

Nordlee J. A., Taylor S. L., Townsend J. A., Thomas L. A. and Bush R. K. (1996) Identification of a Brazil-nut allergen in transgenic soybeans. *N Engl J Med* 334, 688-92.

Oezguen N., Zhou B., Negi S. S., Ivanciuc O., Schein C. H., Labesse G. and Braun W. (2008) Comprehensive 3D-modeling of allergenic proteins and amino acid composition of potential conformational IgE epitopes. *Mol Immunol* 45, 3740-7.

Radauer C. and Breiteneder H. (2006) Pollen allergens are restricted to few protein families and show distinct patterns of species distribution. *J Allergy Clin Immunol* 117, 141-7.

Radauer C. and Breiteneder H. (2007) Evolutionary biology of plant food allergens. *J Allergy Clin Immunol* 120, 518-25.

Roy D., Ghosh D. and Gupta-Bhattacharya S. (2003) Homology modeling of allergenic cyclophilins: IgE-binding site and structural basis of cross-reactivity. *Biochem Biophys Res Commun* 307, 422-9.

Saitou N. and Nei M. (1987) The neighbor-joining method: a new method for reconstructing phylogenetic trees. *Mol Biol Evol* 4, 406-25.

Sali A. and Blundell T. L. (1993) Comparative protein modelling by satisfaction of spatial restraints. *J Mol Biol* 234, 779-815.

Shin D. S., Compadre C. M., Maleki S. J., Kopper R. A., Sampson H., Huang S. K., Burks A. W. and Bannon G. A. (1998) Biochemical and structural analysis of the IgE binding sites on ara h1, an abundant and highly allergenic peanut protein. *J Biol Chem* 273, 13753-9.

Silvanovich A., Bannon G. and McClain S. (2009) The use of E-scores to determine the quality of protein alignments. *Regul Toxicol Pharmacol* 54, S26-31.

Stanley J. S., King N., Burks A. W., Huang S. K., Sampson H., Cockrell G., Helm R. M., West C. M. and Bannon G. A. (1997) Identification and mutational analysis of the immunodominant IgE binding epitopes of the major peanut allergen Ara h 2. *Arch Biochem Biophys* 342, 244-53.

Tamura K., Dudley J., Nei M. and Kumar S. (2007) MEGA4: Molecular Evolutionary Genetics Analysis (MEGA) software version 4.0. *Mol Biol Evol* 24, 1596-9.

Zhang C. T. (1997) Relations of the numbers of protein sequences, families and folds. *Protein Eng* 10, 757-61.

Zuckerkandl E. and Pauling L. (1965) Evolutionary divergence and convergence in proteins. *in Evolving Genes and Proteins, edited by V. Bryson and H.J. Vogel. Academic Press, New York.,* 97-166

2

Family Based Studies in Complex Disorders: The Use of Bioinformatics Software for Data Analysis in Studies on Osteoporosis

Christopher Vidal and Angela Xuereb Anastasi
University of Malta,
Malta

1. Introduction

Complex diseases are common within human populations and communities and pose a great burden not only to affected individuals, but also to society and the health system. Disorders such as chronic heart disease, diabetes, Alzheimer's, epilepsy and many others, are caused by complex interactions of a number of genetic and environmental factors. This makes the identification of the responsible genes difficult if using the same methodologies used for monogenic diseases. For more than fifteen years there has been a collective effort by researchers from around the world to identify genes and genetic variations that increase the risk for osteoporosis and fractures in ageing populations to identify novel therapeutic and prognostic targets, but predominantly most studies have been inconclusive. Genetic heterogeneity between different populations is the main factor responsible for this lack of concordance between different studies. Using different approaches such as association, family linkage, genome-wide association and meta-analysis, researchers reported numerous genes that might play a role in bone physiology, most of the time searching for correlation with phenotypes such as low bone mineral density (BMD) and fractures. Unfortunately, most of these genetic variations were not further investigated for their functional role and how these could lead to the disease. Some monogenic bone diseases led to the identification of genes that were never considered to be involved in bone physiology such as the low density lipoprotein receptor-related protein (LRP)-5 (Gong et al., 2001) and sclerostin (SOST) genes (Brunkow et al., 2001).

A genome-wide linkage scan was performed in two Maltese families with a very high incidence of osteoporosis, where suggestive linkage to chromosome 11p12 was observed. After investigating the genes known to be found at this region by DNA sequencing, we identified a variant in the CD44 gene that was co-segregating with the inherited haplotype in all affected members within one of the families. Further studies on this variant suggested that it could affect pre-messenger RNA splicing, or organisation, leading to different levels of slightly modified variants of the same protein (isoforms). Other loci were identified in both families.

Without doubt, the analysis of data would not have been possible without the number of bioinformatic tools and software that are available. The advances in computer technology including the internet, led to the development of various software and online tools. In this chapter, we will take a look at software and other online tools used in this study. We will discuss the basic concepts of the study, how the analysis was performed using different software and the interpretation of results.

1.1 Gene mapping using families

One of the greatest challenges for geneticists is the identification of genes responsible for complex disease. Unlike classical Mendelian disorders, these diseases do not show obvious Mendelian patterns of inheritance and involve complex interactions between various environmental and genetic factors. Confounding factors such as heterogeneity, phenocopies, genetic imprinting and penetrance further complicate the identification of susceptibility genes. When performing a genetic study, correlation between phenotype and genotype is sought. In complex traits, this correlation might be very low due to incomplete penetrance where not all individuals having the same susceptibility allele are affected or where affected individuals do not have a susceptibility allele (phenocopies). These factors lead a wide ranging severity of disease even within a single family. Further more, late onset diseases such as cardiovascular disease and osteoporosis show up later in life and thus unaffected individuals tested today might become affected in the near future. Late onset diseases are more sensitive to environmental (mostly lifestyle related) factors and are observed to have a higher level of genetic variation due to weak selective pressures on these variants that are usually neutral early in life (Wright et al., 2003). Besides testing for a qualitative trait where individuals are grouped as either having or not having the disease, one can use quantitative or a continuous measurement such as BMD. When using a quantitative variable one must be very cautious as it might not completely correlate with the disease and it could also be dependent on a number of other non-genetic factors including limitations of methodology.
Complex disorders are most often polygenic where multiple genes contribute to the phenotype. Complex patterns of inheritance might be due to allelic or locus heterogeneity where different variants within the same gene are responsible for the disease or where a number of different genes are involved in the same biological process. When studying complex disorders, therefore, one is looking for susceptibility alleles at multiple loci that together increase the individual's risk for the disease. In polygenic traits, penetrance is determined by the genotypes of other loci and therefore it is likely to be low and will vary between individuals. To increase the chance of successful gene mapping, it is important to identify families from probands with extreme phenotypes, earlier age at onset or else to study families from an isolated population with a very high incidence of disease. Wright and colleagues (2003) suggested that it is important to identify genes with the largest contribution to the extremes of the trait and avoid quantitative trait loci (QTL) that have minimal effects on the individual or disease mechanism. Using single extended families from populations that are homogeneous and consanguineous has proven to be a successful approach in localising the genes and novel mutations in type 2 diabetes (Kambouris, 2005). Using one extended family, Kambouris reported similar results to those obtained from previous genome-wide scans using hundreds of individuals (Hanson et al., 1998). This shows that costs and time to identify novel genes responsible for complex disorders can be significantly reduced, when using extended and consanguineous families coming from homogeneous populations.

1.1.1 Linkage analysis

In linkage analysis, the non-independent co-segregation of marker and disease locus is tested in families with multiple affected individuals. Linked alleles (marker with disease-causing allele) on the same chromosome segregate together more often than expected by chance; i.e. against Mendel's law of independent assortment. Gene mapping of a trait identifies chromosomal loci that are shared among affected individuals and that differ between affected and non-affected family members. Positive linkage can only be obtained for marker alleles inherited together with disease allele on the same chromosome. This is a major limitation for linkage analysis when different disease alleles present at the same locus are on different chromosomes, hence *in trans,* as in a case of coeliac disease (Vidal et al., 2009a). In this study, no evidence of linkage was observed to the human major histocompatibility complex (MHC) locus on chromosome 6, in a family with high incidence of celiac disease. Further investigations showed that this was because inherited risk alleles coding for HLA group *DQ2.2* occur *in trans* and so cannot be detected by linkage.

For a linkage study family members from pedigrees with normal and osteoporotic individuals are genotyped for a set of polymorphic markers either across the whole genome or at specific chromosomal loci, where known candidate genes are located. Genetic linkage is measured by the recombination fraction that is the probability that a parent will produce a recombinant offspring and is dependent upon the distance between loci. The more distant two markers are from each other the higher is the chance that a recombination event occurs between them during meiosis. The recombination fraction theta (θ) ranges from 0 for completely linked markers to 0.5 for unlinked loci. Genetic linkage is measured in centiMorgans (cM), where 1cM represents 1% recombination or $\theta = 0.01$ that is equivalent to 1 million base pairs. So using the recombination fraction one can calculate the physical distance on the chromosome, although recombination rates might vary depending on location on chromosome. Recombination rate is usually lower closer to the centromere. Also these measurements might not be so accurate for longer chromosomal distances where multiple crossovers might occur during a single meiotic event, a phenomenon known as interference. Two mapping functions to convert recombination fraction into map distance are Haldane's, that does not assume interference, and Kosambi's, which assumes interference as $1 - 2\theta$.

1.1.2 Parametric linkage analysis

Parametric linkage analysis is a statistical approach using the logarithm of the odds ratio (LOD score) to assess the strength of linkage. This is also known as a model based linkage where the mode of inheritance, frequencies of disease and marker loci together with penetrance must be known. The statistic assumes the likelihood (or probability) that a disease and marker loci in a family are not inherited together ($\theta = 0.5$) compared with the likelihood that they are linked over a selected range of recombination fractions (θ range of 0 to 0.5). The LOD score is the base ten logarithm of the likelihood ratio that is calculated for each value of θ. A two point LOD (z) score (between disease locus and marker) is calculated using the following equation:

$$z(x) = \log_{10} [L(\theta=x) \div L(\theta=0.5)] \quad (1)$$

where x is a value of recombination fraction and L is the likelihood.

Significant evidence of linkage is taken at a LOD score of 3.0 or higher and linkage is completely excluded with a LOD score of -2.5. A LOD score of 3.0 corresponds to odds of 1000:1 that means that it is 1000 times more likely that the alternate hypothesis in favour of linkage holds while a LOD score of 3.5 is equivalent to odds of 3162:1. The LOD score can be converted to a chi-square statistic by simply multiplying by 4.6 and calculating a p-value at

1 degree of freedom (df) for ordinary LOD and at 2 df for heterogeneity LOD scores (HLOD), under the null hypothesis (Ott, 1991). The p-values obtained are always divided by 2 for one-sided tests except when calculating p-values for multi-point LOD (MLOD). Using these calculations a LOD score of 3.0 is equivalent to a p-value of 0.0001 while that of 3.6 is equivalent to 0.00002. However, a chi-square derived p-value applies more for large sample sizes and can be underestimated when sample size is too small. Lander and Kruglyak (1995) suggested that linkage must be reconfirmed by other independent investigators where a nominal p value of 0.01 would be required, while they advised caution when reporting LOD scores that are less than 3.0 and so are only suggestive of linkage. In case of suggestive linkage, additional family data would be required before conclusions can be drawn (Lander & Kruglyak, 1995).

LOD scores can be influenced by a number of factors including the phase or whether parental genotypes are known, misspecification of disease and marker allele frequencies, penetrance, heterogeneity and mostly by phenotypic misclassification. Also for more accurate linkage information and to better localize the disease gene, multi-point linkage analysis is preferred over two-point analysis. Statistical analyses in complex pedigrees are carried out using software such as MLINK and GENEHUNTER where the LOD score can also be adjusted for locus heterogeneity (HLOD) (Kruglyak et al., 1996).

Another kind of analysis which is thought to be useful when analysing linkage data for complex traits is the MOD-score. In complex traits both the genetic model and disease allele frequency are very difficult to specify correctly. An incorrect assumption of the genetic model can significantly affect the analysis and can lead to a false negative result. The MOD score is calculated by maximising the LOD score over a number of replicates using different penetrances and disease allele frequencies, to obtain a maximum LOD score using the best genetic model (Strauch et al., 2003). To control type I errors, it was found that a MOD-score of 3.0 should be adjusted by a value ranging from 0.3 – 1.0 where it was proposed that a MOD-score of 2.5 is indicative of suggestive linkage (Berger et al., 2005). MOD-score analysis can be used to determine the best genetic model for those regions indicated by an initial genome scan using ordinary LODs and it can also be calculated assuming paternal or maternal imprinting. When assuming imprinting a heterozygote paternal penetrance is also used with the other three penetrances with a total of four penetrance values. If a low heterozygote frequency is calculated for paternal imprinting, it indicates that maternal genes are preferentially expressed at that locus (Strauch et al., 2005; Berger et al., 2005).

1.1.3 Non-parametric linkage analysis

Since the mode of inheritance for complex disorders is uncertain, evidence of linkage might be missed by using the LOD score method described above. A more appropriate approach is that described by Kruglyak et al (1996) known as a non-parametric linkage (NPL) or a model free analysis. The NPL statistic measures allele sharing among affected relative pairs (ARP) and/or affected sib-pairs (ASP) within a pedigree. By chance it is expected that siblings share zero, one or two marker alleles identical by descent (IBD) with a probability of 0.25, 0.50 and 0.25, respectively. If disease and marker alleles are linked then affected siblings will share these alleles more frequently than expected by chance regardless of the mode of inheritance. Comparison between expected and observed allele sharing between ASPs is then analysed using the chi-square statistic. Highly heterozygous markers, multipoint linkage and genotyping of non-affected siblings when parents are not available help to increase the sharing information. One great advantage of the NPL statistic is that data from markers on a chromosome can also be evaluated in a multipoint approach using software such as GENEHUNTER which uses the Lander-Green algorithm to calculate IBD distribution (Kruglyak, 1996).

1.2 Phenotype definition, selection of family and population

1.2.1 Phenotype

Phenotype definition is one of the most important factors and should be determined by proper diagnosis or exclusion of other medical conditions that could lead to the same disease. To exclude disease and other factors leading to secondary osteoporosis, individuals were asked to answer a questionnaire and a series of other medical tests were performed. Measurement of bone mineral density (BMD) together with t-scores (number of standard deviations from the mean BMD of a control group of young women at peak bone mass) is the gold standard to diagnose osteoporosis, as recommended by World Health Organisation (WHO). However, this methodology does not show the whole picture partly because bone strength, thus fracture risk, is not completely assessed by measuring bone density. Also, individuals with normal BMD, who might become osteoporotic in ten or twenty years time, could still carry the responsible allele. As discussed above, miss-classification of affected status might seriously affect the results obtained by statistical analysis. To overcome this issue, and unlike other linkage studies for osteoporosis, we used different thresholds of t-scores and z-scores at the lumbar and femoral sites obtained after measuring BMD, to define discreet phenotypes as simply affected or not-affected. Statistical analyses were performed in five different scenarios defining discreet phenotypes using the guidelines suggested by the International Society of Clinical Densitometry (Khan et al., 2004).

1.2.2 Families

Extended families with a number of affected individuals are ideal for identifying variants with higher penetrance but are less frequently found in populations. Development of novel treatments can be targeted to these pathways. Factors such as mode of inheritance, penetrance and disease or allele frequencies together with technical factors such as accuracy of genotyping, all affect power to detect a significant linkage.

1.2.3 Population

The genetic component within a population is strongly affected by its history and demography. The genetic pool of a population is determined by mutations, population admixture as well as by random genetic drift that occurs most often due to catastrophic events that result in a major decrease in population (Wright et al., 1999). Genetically isolated populations (by geography and/or culture), that recently expanded from a very small number of founders with occasional interbreeding with other ethnic groups, are more likely to share haplotypes identical by descent (IBD) over longer genetic distances (Wright et al., 1999).

The present Maltese population, although geographically (but not genetically) isolated, is thought to have expanded exponentially from a much smaller population during the last four hundred years with a possibility of a number of founder effects introduced by admixture with other populations coming from Sicily, the eastern Mediterranean and northern Africa. Founder effects were reported in the Maltese population, including a mutation (R1160X) found in the NPHS1 gene coding for nephrin that causes nephrotic syndrome (Koziell et al., 2002) and the 68G>A mutation within the quinoid dihydropteridine reductase gene that causes a rare form of hyperphenylalaninaemia and phenylketonuria (Farrugia et al., 2007). The introduction of founder effects and major bottlenecks may increase the chance of creating sub-populations with particularly high allele frequencies when compared to the rest of the population (Heiman, 2005). Significant fluctuations in the population were brought about by emigration of the Maltese in fear of further attacks by the Turks, death by famine or plague. On the other hand, the existence of a relatively frequent disease in an island population does not necessarily always indicate a possible founder effect since this might result from multiple mutations in a single gene or in different genes that could lead to the same phenotype (Zlotogora, 2007).

Genetically isolated populations proved to be very useful for the identification of genes not only in the case of the BMP-2 gene in Iceland but also for a number of other diseases (Styrkarsdottir et al., 2003). More than 15 mutated genes were successfully identified by positional cloning in families from the isolated population of Finland. The Finnish population demographic history was characterised by rapid expansion from a much smaller population with a number of founder effects (Peltonen, 2000). Another island population that proved successful for the identification of a mutation responsible for uric acid nephrolithiasis by linkage was the Sardinian population (Gianfrancesco et al., 2003). Linkage studies in Maltese families resulted in successful identification of rare genetic variants responsible for other human disorders such as coeliac disease (Vidal et al., 2009a), epilepsy (Cassar, 2008) and recently in the identification and confirmation of the role played by the erythroid transcriptional factor KLF1 in hereditary persistence of foetal haemoglobin (Borg et al., 2010).

2. Materials and methods

2.1 Patient recruitment

Two extended families consisting of a total of 27 family members with several individuals having low BMD were recruited for this study. Families were selected through index patients (or probands) referred to the Bone Density Unit, Department of Obstetrics and Gynaecology, St. Luke's Hospital, Malta for an osteoporosis risk evaluation. The proband in Family 1 was a 61-year-old female diagnosed with osteoporosis six years earlier and was known to have a family history of osteoporosis. Five out of seven of her siblings were recruited while the other two were not willing to participate in the study. Osteoporosis was confirmed in all six recruited siblings. All female siblings were osteoporotic at the lumbar spine and one male was osteoporotic at the femoral neck. One sibling had an asymptomatic compressed vertebral fracture. Three daughters of the proband were recruited (age range 33 - 38 years) and all of them were found to have very low BMD for their relatively young age. Their 37-year-old cousin was also found to have very low BMD at both the lumbar (t-score -2.25) and femoral neck (t-score -1.07), and had very low body mass index (BMI) (16.2 kg/m^2). It was not possible to collect blood for DNA analysis from this participant.

The proband in Family 2 was a 55-year-old woman with osteoporosis at the lumbar spine, diagnosed five years earlier. A closer investigation of this family revealed four osteoporotic siblings out of five. Their children were healthy young adults, some of whom had very low BMD relative to their age. The presence of males with low BMD and history of fractures in a severely osteoporotic sibling were good indicators that a genetic factor might be involved.

As already discussed, five different scenarios were tested using thresholds for t-scores and z-scores as previously described (Khan et al., 2004). Osteoporosis for post-menopausal women and men over fifty years of age was defined using a lumbar and/or femoral t-score of less than -2.50 (WHO criteria). Definition of affected status for younger individuals was determined using z-scores of less than -1.0 and less than -2.0 for a more severe phenotype, for scenarios III and IV, respectively. For scenario V, analysis was performed using only affected individuals having femoral z-scores of less than -1.0. In all five scenarios, family members having normal BMD measurements were assumed to have an unknown phenotype. This assumption takes into consideration the possibility that any apparently clinically unaffected individual might actually be affected, thus reducing the chance of obtaining false negative results.

2.2 Genotyping

To perform a successful gene mapping study, a number of polymorphic markers have to be typed in affected and non-affected individuals to identify genes that increase the risk of disease. Different types of genotyping markers were used in recent years and new techniques for typing are constantly being developed to increase efficiency, accuracy and throughput while reducing costs.

2.2.1 Microsatellite genotyping

Short tandem repeats (STRs) or microsatellites are widely distributed in the genome and so are useful tools for genome-wide scans. These tandem repeats can be dinucleotide, trinucleotide or tetranucleotide repeats where polymorphisms are generated by gain or loss of repeats usually as a result of both replication slippage and point mutation. Microsatellites have several advantages for typing, the most important of which is that they are highly polymorphic with a very high heterozygosity (>70%), so making them ideal for use in linkage studies. Another advantage is that they can be very easily typed using PCR techniques where fluorescently labelled primers flanking the polymorphic region are designed. The variable number of repeats creates amplicons of different sizes which can be typed using automated sequencers such as those by Applied Biosystems (ABI) (PE Applied Biosystems Division, Foster City, CA). Different sets of markers across the whole genome are electronically available from databases such as those of Marshfield Institute of Genetics (http://research.marshfieldclinic.org/genetics/), deCode (http://www.decode.com/services/microsatellite-genotyping-genome-wide-scans.php) and the Cooperative Human Linkage Centre (http://gai.nci.nih.gov/CHLC/). Markers can be selected from these databases either across the whole genome or at candidate loci usually with an average spacing of 10cM and for a higher resolution at < 5cM. To increase throughput and reduce costs, the amplified fragments are carefully pooled in sets in such a way that the allele size range does not overlap within a set and by using different dyes for different sets.

An initial genome-wide scan, 400 microsatellite markers spread across the 22 autosomes and x-chromosome with an average spacing of 8.63cM and heterozygosity of 0.77, was performed. The average performance of markers for all samples was of 96.96%. Fine-mapping was performed by increasing the markers at indicated loci from the initial scan. Genotyping was performed by polymerase chain reaction (PCR) followed by fragment analysis using a 3730xl ABI genetic analyser (Applied Biosystems, Foster City, CA, USA). The average performance of the markers was of 96.02%. Genotyping was performed commercially at the McGill University and Genome Quebec Innovation Centre, Quebec, Canada.

2.3 Analysis of linkage data

PedCheck (O'Connell and Weeks, 1998) was used to determine if the inheritance of marker loci was according to Mendel's laws. Multipoint parametric and non-parametric linkage analyses were performed using GENEHUNTER-PLUS (Markianos et al., 2001) which is an improved version of GENEHUNTER (Kruglyak et al., 1996). GENEHUNTER v1.2 was used to calculate Zlr scores according to Kong and Cox (1997). Linkage analysis of markers on the X-chromosome was performed using a specific application for this chromosome included with the GENEHUNTER package. All analyses were performed using EasyLinkage v5.05 (http://www.uni-wuerzburg.de/nephrologie/molecular_genetics/molecular_genetics.htm) (Lindner and Hoffmann, 2005). Parametric analysis was carried out using variable penetrances for both a dominant and recessive mode of inheritance. Penetrances used for the dominant model were 0.01 for the wild-type homozygote, 0.90 for mutant heterozygote and 0.90 for mutant homozygote, respectively. The recessive model was defined by penetrances 0.01, 0.01, and 0.80 for the wild-type homozygote, mutant heterozygote and mutant homozygote, respectively. A more complex model was also analysed using penetrances 0.01, 0.05, 0.30 for wild-type homozygotes, mutant heterozygotes and mutant homozygotes, respectively. A parametric analysis assuming heterogeneity was computed using data from both families (HLOD).

A co-dominant allele frequency algorithm was used for the analysis, as suggested in the EasyLinkage manual, for extended families. For all models, the disease allele frequency assumed was 0.001, and phenocopy rate of 1%. This disease allele frequency is equivalent to a population prevalence of 0.2% assuming Hardy-Weinberg equilibrium calculated using the following equation (Xu & Meyers, 1998):

$$2(1 - q)\, q + q2 \qquad (2)$$

q = disease allele frequency.

Analysis was performed using other penetrance values for loci showing evidence of linkage in the initial genome-wide scan. The exact genetic model was determined using GENEHUNTER-MODSCORE v1.1 (Strauch et al, 2005), where MOD scores were calculated from simulations of different models and disease allele frequencies with and without imprinting. This analysis was suggested by Strauch et al (2003) for complex trait analysis and was done only for those regions showing suggestive linkage. The deCode genetic map was used throughout the study.

2.3.1 Using EasyLinkage v5.05 graphical user interface (GUI)

EasyLinkage is a Microsoft Windows® based GUI, developed in recent years. This was a step forward for researchers wanting to perform linkage analysis. Using EasyLinkage and a common input file format, one can analyse data using all major software such as PedCheck, GENEHUNTER, Merlin and Allegro. EasyLinkage can be used to analyse data generated from projects using both single nucleotide polymorphisms (SNPs) as well as STRs. Analysis can be performed on chromosome by chromosome or else genome-wide basis, making use of the appropriate genetic maps (such as deCode and Marshfield), using male, female or sex-averaged maps, from which more accurate genetic positions can be drawn. Both graphical and text output files are automatically generated for each individual family together with a collective report averaging all families, in text or pdf formats and stored into an appropriately labelled folder showing date and type of analysis. These files show statistical analyses results such as LOD scores, NPL, p-values and input parameters given by the user for that model including penetrances, disease allele frequencies and genetic positions of markers ranked according to the most significant results.

There are four allele frequency algorithms to choose from depending on the type of analysis needed. Several versions of this software have been developed, improving its capabilities to handle large amounts of data generated from SNP arrays such as the Affymetrix 500k and Illumina 650k chips. For SNP analysis, allele frequencies of all the major ethnic groups form part of the EasyLinkage software package.

2.3.2 Data entry

There are two main types of files needed to perform linkage analysis using microsatellites or STRs. In this study, a qualitative type of analysis was performed using discrete phenotypes (affected vs unaffected). One type of input file should contain family or families' information in a standard linkage format. The marker file should include the genotype results for each family member. All family or families' information including pedigree structure has to be entered into a pedigree file. Shown below is part of the pedigree file as created in our study (only obligatory columns were used). From left to right columns represents (i) unique family identifier; (ii) individual unique identifier (iii) father and (iv) mother identifiers; (v) sex identification code (1=male, 2=female, 0=unknown); (vi) affected status (1=unaffected, 2=affected, 0=unknown). In case parents are unknown then enter '0`. As explained previously is an unknown phase and so it reduces the power of the study, even though the software is able to assume the genotypes of these individuals using the known genotypes from their offspring (inferred genotypes). An example for using the unknown option in column (v) is when you do not know the sex of a child due to death *in utero*.

A_1	A_1_01	A_1_11	A_1_12	2	2
A_1	A_1_02	A_1_11	A_1_12	1	2
A_1	A_1_03	A_1_11	A_1_12	2	2
A_1	A_1_10	0	0	1	0
A_1	A_1_11	0	0	1	0
A_2	A_2_11	A_2_28	A_2_29	1	1
A_2	A_2_12	A_2_28	A_2_29	2	2

Phenotype definition has to be done using appropriate criteria and diagnostic tests, for example in our study, measurements of BMD together with blood tests were used to exclude other medical conditions that could also affect BMD. In complex disorders, it might be difficult to define the phenotype correctly and this could seriously affect the outcome of results. Select 1 and 2 wherever diagnostic tests were performed and phenotype is known. Any individuals not tested should be defined as having an unknown status. This is a better option because individuals wrongly defined as normal could give a false negative result (type II error) as these might be carrying the causative alleles and might become affected at a later stage in their life. As described in the phenotype definition section, we analysed our data using five different scenarios defined by t-scores and z-scores. For each scenario a different pedigree file was created and saved in a folder together with the marker files described in the next section.

Creating marker files

Marker files should include genotyping results for all family members tested. Entering data is the most laborious part of the study because different files have to be created for each marker, i.e. if 400 markers were tested then 400 different marker files have to be created and saved in the same folder together with pedigree files. These files should be named with the marker identification corresponding with that in the marker map file (.map) used by the software e.g. 'D1S200_FINAL.abi`. If the marker is not found within the marker map file then an error is given when running the analysis. This error can be corrected manually by adding the marker into the marker map file found in the EasyLinkage folder in Program Files.
This is an example of the method used to input data into marker files:

MARKER	LANE	ID	A_1	A_2
D1S200	A1	A_1_01	165	176
D1S200	A2	A_1_02	161	176
D1S200	A3	A_1_03	165	176
D1S200	A4	A_1_04	161	176
D1S200	A5	A_1_05	161	176
D1S200	A6	A_1_06	161	176
D1S200	B1	A_2_10	161	176
D1S200	B2	A_2_11	161	176
D1S200	B3	A_2_12	161	161

Column (i) name of marker e.g. D1S200; (ii) PCR reaction position in a 96-well PCR plate (information not used by software); (iii) individual identification number corresponding with pedigree file; (iv) allele 1 in base pairs (bp) and (v) allele 2 in bp. Any missing genotypes should be entered as '0`. When analysing data, the software will re-code these alleles numbering them consecutively as 1, 2 etc depending on the number of alleles observed for that marker in all genotyped individuals. The higher the number of alleles observed the higher the heterozygosity and thus the more informative that marker is.

2.3.3 Running EasyLinkage analysis

On the main screen of the GUI, we selected a 'Single Locus` analysis, the linkage software (GENEHUNTER) and microsatellites project type. Next step was to select whether to perform a genome-wide analysis, one chromosome at a time or even to analyse small segments from a chromosome. Analysing small segments from a chromosome is useful to analyse large scale SNP data possibly analysing 500 markers in one segment. A lower LOD score was observed when analyzing a large number of markers, which would mean that for SNP analysis, it is better to avoid SNPs that are very close to each other. LOD scores were observed to be lower in such instances most likely due to allele frequencies used. It would be advisable to first analyse the whole chromosome for SNP analysis, and if significant results are observed, then re-analyse blocks of 100 markers at a time and as overlapping blocks. Another suggestion would be to use different and appropriate allele frequency algorithms, as will be described below.

After choosing the chromosomes, the sex-averaged deCode genetic map was selected to position the markers. There is a difference of approximately 10 cM between the male and female genetic maps, being longer in females due to a higher recombination rate. Other general options selected included 'recode alleles` for continuous recoding of alleles within the marker files, Mendelian testing using PedCheck and the autoscale Y-axis for LOD / NPL plots.

Finally we chose the folder where the pedigree files were saved and the option to give individual pedigree results as well as totals. As decribed earlier, five different phenotype scenarios were used and each one had to be analysed using a different pedigree file.

GENEHUNTER (GH)

This computer package was developed to perform multipoint linkage analysis (parametric and non-parametric) in pedigrees of moderate size (Kruglyak et al., 1996). The program can compute LOD scores for pedigrees using a mode of inheritance and penetrance specified by user. It also allows the user to test for linkage under genetic heterogeneity. The multipoint NPL analysis tests for IBD allele sharing among affected individuals within pedigrees that is not affected by the mode of inheritance. It is thus ideal to be used for complex traits. GH also constructs the most likely haplotypes indicating crossovers even if there is missing data. A major advantage of GH over other statistical software, such as VITESSE and MLINK, is that it uses the Lander-Green algorithm and therefore it can perform multipoint analysis using several markers on a chromosome. Major drawbacks of GENEHUNTER include restrictions on pedigree size and its relatively slow speed when compared to similar software such as Allegro. Another limitation of GENEHUNTER is that it cannot analyse large number of markers which means that if one was analysing more than 100 markers on same chromosome, one would have to analyse these in groups of 100, repeating the analysis with different set sizes so as not to miss the signal. A recent version of GH can also perform transmission disequilibrium testing (TDT) analysis and analysis of quantitative traits making GH the ideal software to use for genetic analysis (Nyholt, 2001). In this study GENEHUNTER v1.2 was used to calculate Zlr scores using the Kong and Cox (1997) model. This algorithm addresses the problem encountered by previous versions of GENEHUNTER where NPL scores were found to be too conservative when inheritance data was incomplete. Another application used in this study was GENEHUNTER-MODSCORE v1.1 that maximises LOD scores with a series of penetrances and disease allele frequencies (Strauch et al., 2005).

Using GENEHUNTER (GH) with EasyLinkage GUI

Performing linkage analysis using GH through the Easylinkage GUI is easy and straightforward and saves time. After choosing the GENEHUNTER package software as described above, one has to go to Program Options from the main dashboard to be able to define a model for analysis. We analysed our data using both dominant and recessive models of disease. GENEHUNTEr v1.2 and GENEHUNTER-MODSCORE v1.1 were used for the analysis. A 'Codominant` allele frequency algorithm was used for our analysis. EasyLinkage gives you four different algorithms to choose from. The Codominant algorithm was the best choice to use for extended families. This algorithm uses only alleles from genotyped individuals within the pedigree file. Frequencies of the alleles are calculated to sum up to 1, which means that if 5 different alleles were observed, then the allele frequency for each allele will be set to 0.200 or if 10 then to 0.100. If less than 5 alleles are found then still the frequency is set to 0.200. Other allele frequency algorithms include either all

individuals within the marker file or all individuals from pedigree file, both suitable for the affected sib pair (ASP) design. There are also specialised algorithms such as 'founders only`, suitable only for pedigrees with large number of founders. SNP projects will use reference allele frequencies from different ethnic groups.
As described before, we analysed our data using variable penetrances of disease starting with a highly pentetrant form (90%) down to 50%, for each scenario. Disease allele frequency was taken as 0.001 and the analysis steps between markers for multi-point analysis were set to 5, with recombination counting set to 'On`. Penetrances were entered into the appropriate fields as described before, turning the haplotyping options to 'on` and choosing the 'Display all family plots`. The haplotyping option significantly increases the run time of the analysis but it creates plots for each family with haplotypes and marker positions together with other files that can be used by other software such as HaploPainter (Thiele & Nurnberg, 2004).

3. Results

3.1 Reading analysis files

All analysis files were saved into an appropriately labelled folder with details of software used, allele frequency algorithm, pedigree file name, date and time of analysis. Data was saved as text and post-script format. The '.OUT` file within the 'LOG` folder can be opened using Notepad where one can find all commands given to GENEHUNTER by EasyLinkage. If any errors were encountered and analysis was not completed, then one would find all information logged in, within this file. This file also includes detailed results such as LOD / NPL scores, marker information etc for each individual family and totals for all the families. EasyLinkage automatically commands GENEHUNTER to set up the maximum number of bits to be analysed within the family and to 'trim` large families when needed. These two functions are needed to keep computations within the computer running ability due to memory limitations. Max Bits function is dependent on the number of meiosis being examined and represents 2N – F, where N is the number of non-founders and F is the number of founders. If, in the family there are 10 children (non-founders) from 2 sets of parents (4 founders), then max bits will be set to 16. If the family is larger, then the less informative individuals will be excluded from the analysis using the 'trim` command. Automatically EasyLinkage also commands GENEHUNTER to use the Haldane map and there is no option to use the Kosambi map. The Haldane map does not assume interference, as described in a previous section.
Figure 1A, shows the upper left hand quandrant of a parametric LOD score plot, including information such as allele frequency algorithm used, inheritance model, allele frequencies and penetrancies expressed as percentages. Markers are ranked in order of their most significant LOD / NPL scores with information about chromosome number, their position on the genetic map and calculated probability (Figure 1B). Figure 1C shows NPL results for a genome-wide analysis with NPL scores on the y-axis and chromosomal marker positions on the x-axis. Other plots are given for LOD scores, HLODs, p-values, Zlr scores and marker information.
If the haplotyping option was chosen then a folder named 'Haplotyping` is created and haplotype files for each pedigree are saved there. This haplotype data can then be imported to other software such as HaploPainter (Thiele & Nurnberg, 2004) to construct a graphical representation of the inherited haplotypes.

A

Project:	OSTEOPOROSIS	**Inheritance:**	Dominant
Family name:	A_1	**Common allele:**	99.90 %
Used map:	decode strp[1] (sex averaged)	**Disease allele:**	0.10 %
Marker positions:	24 ok / 0 ? / 0 outside	**LC1 PCOPY rate:**	1.00 %
Allele frequencies:	Equal distribution of alleles	**LC1 PENET wt/mt:**	90.00 %
HAP algo / SCFct:	On / All	**LC1 PENET mt/mt:**	90.00 %

B

Marker	CHR	cM	pLOD	NPL	npLOD	P val	INFO	Alpha	hLOD
1. D6S309	6	20.50	0.4256	−0.3697	0.0297	0.7188	0.9222	-	-
2. D6S1006	6	30.18	0.4235	−0.3736	0.0303	0.7188	0.9201	-	-
3. D6S309 – D6S1006	6	22.44	0.4230	−0.3694	0.0296	0.7188	0.9026	-	-
4. D6S1713 – D6S309	6	18.66	0.4229	−0.2391	0.0124	0.4375	0.6361	-	-
5. D6S309 – D6S1006	6	28.24	0.4217	−0.3717	0.0300	0.7188	0.9016	-	-

C

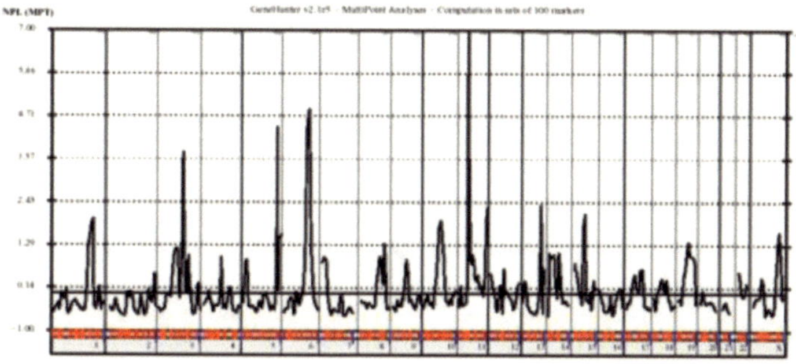

Fig. 1. Information given by parametric LOD score analysis plots (A & B) and genome wide NPL plot (C)

3.2 Linkage results

From the initial genome-wide scan using both families, evidence of linkage was observed to marker D11S1392, where the highest NPL score was of 5.77 (p=0.0006) and LOD/HLOD of 2.55, for the dominant model with 90% penetrance and phenocopy rate of 1%. Fine mapping was performed by analysing four additional markers at this region (D11S4101, D11S935, D11S4102, and D11S1911) with average spacing of 1 to 1.5cM. Fine mapping confirmed linkage to marker D11S935 that is 52.94cM from 11p-telomere. Table 1 shows the highest scores obtained for this marker using the dominant mode of inheritance with 90% penetrance and phenocopy 1%. HLODs are calculated when more than one family are analysed together and thus the score can be different from the LOD if there is heterogeneity between families.

Table 2 shows results obtained when analysing the same families assuming clinically unaffected individuals, whose BMD was measured by DEXA, as normal phenotype (according to WHO criteria) rather than having unknown phenotype (as in Table 1). NPL and Zlr scores were the same as observed in Table 1, but LOD and HLOD scores were different.

LOD scores shown are for the autosomal dominant model with 90% penetrance and 1% phenocopy rate.

Phenotype	LOD (cM)	HLOD (α)	NPL (p-val)	Zlr
Scenario I	2.90 (52.94)	2.90 (1.00)	7.00 (0.0014)	3.01
Scenario II	2.46 (52.94)	2.46 (1.00)	4.02 (0.0038)	2.90
Scenario III	2.89 (52.94)	2.89 (1.00)	7.23 (0.0013)	3.04
Scenario IV	3.35 (52.94)	3.35 (1.00)	6.90 (0.0002)	3.74
Scenario V	2.59 (52.94)	2.59 (1.00)	5.37 (0.0020)	3.28

deCode map position in brackets in cM

Table 1. Highest scores for marker D11S935 in both Pedigrees using an Autosomal Dominant Model after Fine Mapping

This is because in the second analysis, shown in Table 2, based on current BMD measurements were defined as normal, individuals that might be osteoporotic in the future. These individuals might also be carrying the inherited causative allele and so will result in a false negative result if taken as normal. This is a common situation with complex and late onset disorders such as osteoporosis.

Phenotype	LOD (cM)	HLOD (α)	NPL (p-val)	Zlr
Scenario I	3.07 (52.94)	3.07 (1.00)	7.00 (0.0014)	3.01
Scenario II	-0.19 (52.94)	-0.00 (0.00)	4.02 (0.0038)	2.90
Scenario III	2.97 (52.94)	2.97 (1.00)	7.23 (0.0013)	3.04
Scenario IV	2.80 (52.94)	2.80 (1.00)	6.90 (0.0002)	3.74
Scenario V	1.26 (50.64)	1.26 (1.00)	5.24 (0.0020)	3.28

Table 2. Analysis of Chromosome 11 in Both Families Assuming Unaffected Individuals as Normal

When calculating MOD scores for chromosome 11, the highest MOD was of 3.28 at the same region 52.94cM using the best calculated genetic model with penetrances 0.06 wild-type homozygotes (6% phenocopy rate), 0.97 for both heterozygotes and mutant homozygotes. The disease allele frequency calculated at this model was of 0.000006 with a calculated population prevalence of 0.001%, assuming Hardy-Weinberg equilibrium. A MOD score of 4.33 (info = 0.87) was obtained when assuming imprinting with a disease allele frequency of 1×10^{-6}. Estimated penetrances of wild-type homozygote (f +/+) 0.00; paternal heterozygote (f m/+) 0.00; and 1.00 for both maternal heterozygote (f +/m) and mutant homozygotes (f m/m) show evidence of paternal imprinting at this locus. Paternal imprinting indicates that the expression of the gene responsible for the disease at this locus may be entirely maternal.

This locus was further analysed by varying the penetrance and phenocopy rates for the dominant mode of inheritance. Analyses were performed using phenocopies from 1% to 20% and penetrance 0.7 - 0.5. The phenocopy rate is the percentage of individuals within the family that are clinically affected but do not carry the disease allele and hence their phenotype is due to other mainly environmental factors. As shown in Table 3, the highest LOD/HLOD score (3.32) was observed at penetrance of 0.8 and 0.7 with a 5% phenocopy. Changing the penetrances and hence the model, does not affect NPL scores (since these are model free) and therefore NPL scores are not shown in Table 3.

	Family 1			*Family 2*		
	*LOD (cM)**	*NPL (p-val)*	*Zlr*	*LOD (cM)**	*NPL (p-val)*	*Zlr*
Scenario I	1.92 (54.35)	6.26 (0.0078)	2.84	1.04 (52.48)	4.42 (0.0098)	2.41
Scenario II	1.35 (55.77)	3.10 (0.0313)	2.12	1.18 (52.94)	3.06 (0.0625)	2.11
Scenario III	1.92 (54.35)	6.26 (0.0078)	2.84	1.04 (51.56)	4.74 (0.0117)	2.27
Scenario IV	1.64 (54.35)	4.41 (0.0156)	2.58	1.77 (52.94)	5.85 (0.0156)	2.75
Scenario V	0.86 (48.21)	1.94 (0.1250)	1.63	1.75 (50.64)	5.83 (0.0156)	2.88

Table 3. Multipoint LOD/HLOD Scores on Chromosome 11 under an Autosomal Dominant Model with Variable Penetrance and Phenocopy

Although both families shared the same linkage interval, the highest LOD scores were obtained by two different markers with a spacing of approximately 4cM between them, showing also different inherited alleles, suggesting that different genes at the same locus, and within the same linkage interval, might be responsible for the same disease in different families (allelic heterogeneity). Highest LOD and NPL scores (1.77 and 5.9, respectively) were obtained for marker D11S1392 (50.64cM) for Family 2, while for Family 1 highest scores were obtained to marker D11S4102 (54cM) (Table 4). Inherited haplotypes identical by descent were observed in both individual families between markers D11S1392 and D11S935, with a number of recombination events defining boundaries for the linkage interval where the causative genes can be found in between.

Penetrance	**LOD/HLOD Phenocopy = 1%**	**LOD/HLOD Phenocopy = 5%**	**LOD/HLOD Phenocopy = 10%**	**LOD/HLOD Phenocopy = 15%**	**LOD/HLOD Phenocopy = 20%**
0.9	3.07	3.25	3.05	2.78	2.47
0.8	3.10	3.32	3.11	2.82	2.44
0.7	3.12	3.32	3.08	2.73	2.23
0.6	3.14	3.30	2.99	2.54	1.80
0.5	3.17	3.25	2.84	2.17	1.07

Table 4. Analysis of chromosome 11 for Families 1 & 2 after fine mapping * deCode map position in brackets in cM

3.3 Choosing and sequencing candidate genes

The locus indicated by both parametric and non-parametric linkage analyses on chromosomes 11p12 was scanned for known candidate genes. Candidate genes within the linkage interval were selected with prior knowledge of physiology using the NCBI map viewer (http://www.ncbi.nlm.nih.gov/mapview/) *Homo sapiens* build 36. The online application GeneSeeker v2.0 (http://www.cmbi.kun.nl/GeneSeeker) was also used. A new online tool GeneDistiller (http://www.genedistiller.org/) was recently developed to filter genes within a specified linkage interval is (Seelow et al., 2008). When using NCBI

MapViewer to select the genes manually, one has to align the genes with the corresponding genetic map (e.g. deCode), setting the resolution of the map to 1 cM for accurate alignment. Applications such as GeneDistiller can facilitate this process since they automatically extract all genes within a given interval. It is also advisable to search for genes further away from both ends of the linkage interval even up to 5 - 10 cM. This will compensate for differences in positioning of markers on the genetic map and the actual physical map.

The whole area from 49 to 55cM on chromosome 11 (deCode genetic map) was searched for genes that might plausibly be involved in the disease. More than twenty genes and hypothetical proteins are found in this region, with the best candidates being the tumour necrosis factor receptor-associated factor 6 (TRAF6) gene [MIM 602355] and the CD44 gene [MIM 107269] (sequenced in Family 2) found 1cM away from D11S1392 (~51cM). TRAF6 was sequenced in both families but it was closer to D11S4102 showing highest scores in Family 1.

Oligonucleotide primers were designed using the online application Primer 3 (http://frodo.wi.mit.edu/primer3/) (Rozen and Skaletsky, 2000) to amplify all coding regions including intron-exon boundaries and promoter region of the CD44 gene using transcript ENST00000278385) from the ENSEMBL database (http://www.ensembl.org). Transcript ENST00000313105 was used for TRAF6 gene. Due to limitations of the sequencing technique, only fragments from 200 to 600bp were amplified by PCR. Large exons and up to 1500bp of the 5` untranslated region were covered by overlapping PCR fragments. Bidirectional DNA sequencing was performed using standard techniques and fluorescent capillary electrophoresis.

3.4 Reading and Interpretation of sequencing results

DNA sequencing results were compared to reference sequences in public databases by using the software ChromasPro v1.33 (http://www.technelysium.com.au) that directly searches the BLAST application on NCBI. Variations that did not agree with the reference sequence were confirmed by the reverse sequence. Electropherograms were also printed and checked manually.

Detailed information about specific genes including information about known mutations/polymorphisms and gene expression was obtained from GENECARDS (http://www.genecards.org) and The Human Gene Mutation Database (http://www.hgmd.cf.ac.uk/). When identifying a variation, the first step is to check using databases whether it is already known. Information about individual SNPs can be searched in gene and SNP databases such as the NCBI SNP database (http://www.ncbi.nlm.nih.gov/projects/SNP/). A list of SNP databases can be found at http://www.humgen.nl/SNP_databases.html. If the variation is a known SNP, then one has to refer to it using the database reference number such as 'rs3830511`. In the database one can find information about individual SNPs including any population frequencies. The next step is to identify the frequency of these variations in the local population and determine whether it is a rare or common variation. To perform population screening, one can use techniques such as restriction fragment length polymorphism (RFLP), real-time PCR or direct DNA sequencing in a random sample from the general population. For our studies, random samples of DNA were obtained from newborns and used for this purpose, followed by a small scale case-control study using osteoporotic and normal post-menopausal women.

Virtual restriction fragment length polymorphism (RFLP) gel electrophoresis was carried out using the online web applications NEBcutter V2.0 (http:// tools.neb.com /NEBcutter2/index.php) to test the identified polymorphisms.

3.4.1 TRAF6 Sequencing and functional assays

Following direct sequencing, three different variants were identified when compared to reference sequences on the NCBI and Ensembl databases. An A to T transversion was identified at position -721 (5` upstream of exon 1), when compared to TRAF6 reference sequence (AY228337). This variant had not been previously described. Following sequencing of all family members, three affected individuals from Family 1 were observed to be heterozygous for this variant. Individuals from Family 2 were all wild-type homozygotes. RFLP was carried out in 82 unrelated postmenopausal women. This variant was observed to be very rare within an unrelated group of postmenopausal women, as only three heterozygotes were observed. After screening 350 chromosomes in a random sample from the general population, only 2 alleles were observed (0.57%) with this variant having a population frequency of 1.1%.

A previously described insertion/deletion of a T in the intron between exons 4 and 5, in the polyT region, sixteen base pairs ahead of the exon-intron boundary (rs3830511), was also identified. When analysing all family members and controls, only three individuals were observed to be heterozygotes for this insertion/deletion, one of whom was severely affected and the other two were normal individuals.

A transition from G to A was found in the intron between exons 6 and 7, 110bp upstream of the exon-intron boundary. When sequencing all members from both families, four heterozygotes for this variant were identified, and the rest were homozygous for the wild type allele G. Three of the four heterozygous individuals for this variant had a low BMD, two of whom were also heterozygous for the T insertion/deletion described above. Genotyping by RFLP (PvuII) was performed in 82 unrelated postmenopausal women. Genotype frequencies observed were 72.3% GG, 26.5% GA and 1.2% AA.

Although the -721 A/T polymorphism was not linked to the inherited haplotype within Family 1, this polymorphism was rare within the population and it was thus hypothesised that it could affect gene expression. The TRAF6 gene plays a major role in osteoclast differentiation and activation and plays an important role in osteoimmunology (ref).To test this hypothesis the TRAF6 gene promoter region, harbouring the -721 variant, was analysed for possible transcriptional factor binding sites in the presence and absence of the variant identified in this study, using the online application MatInspector by Genomatix Software GmbH(http://www.genomatix.de/online_help/help_matinspector/ matinspector_help.html) (Cartharius et al., 2005). The whole sequence, up to 1500bp upstream from the transcriptional start, site was thus copied and tested using the MatInspector online application. Free registration was needed to use this application for academic use allowing twenty analyses per month. Both normal and mutated sequences were used and analysis was performed using a vertebrate matrix. When comparing normal and mutated alleles it was observed that position -721 might be a potential binding site for nuclear factor Y (NF-Y), a CCAAT binding factor, that binds to the wild-type allele but not to the mutated one. Non-binding of this factor would result in a decreased expression of the gene.

Three different sized fragments from the promoter region of TRAF6 (up to 1500bp) were cloned into a luciferase reporter vector and transfected into two types of mammalian cells. After measuring luciferase activity in both cell lines, it was evident that gene expression was affected by the -721 variant found in the TRAF6 gene promoter. Expression of the mutated allele was observed to be as low as 5% that of the normal allele expressed in murine macrophages. The two longer constructs showed higher expression for the mutated allele suggesting that other transcriptional factors most likely interact either directly or with other factors at the mutated site. Although these observations suggest that this variant affects a transcriptional factor binding site and thus could increase the risk for osteoporosis, further research is needed to identify the molecular mechanisms.

3.4.2 CD44 gene sequencing in family 2

DNA sequencing of the CD44 gene found on chromosome 11p12 was performed in Family 2 since this gene is found closer to D11S1392, which shows the highest LOD scores within this family, as described above. Osteoclast formation was inhibited by CD44 antibody suggesting its important role in bone physiology and as a potential therapeutic target for metabolic bone disease (Kania et al., 1997). As well, CD44 was also associated with inflammatory bone loss (Hayer et al., 2005).

Sequencing CD44 revealed a number of intronic sequence variants, including two A/G changes (rs4756196 and rs3736812) and an A/C transversion in intron 16, none of which were observed to be inherited with the linked marker. A number of other variants were found in coding regions, including an A/G (rs9666607) and C/T (rs11607491) changes in exon 10, both resulting in an amino acid change, which were not linked with the inherited haplotype. Another C/T synonymous variant (rs35356320) was detected in three affected and one unaffected individual. A non-synonymous variant found in exon 12 (rs1467558) was only found in two affected members of this family.

An interesting variant was detected in exon 9, a synonymous G/A transition (rs11033026), found 32 nucleotides upstream from the exon/intron junction. Sequencing the gene in all members of this family, revealed that all individuals carrying the linked STR allele 3 (Figure 1B) for marker D11S1392 were also heterozygous for this variant, suggesting that the two were linked.

As shown in Figure 1B, all affected members, with the exception of one phenocopy (III:4), were heterozygous for both the STR allele and the A allele. This variant was not found in any of the non-affected family members, with the exception of two who also carried the linked STR allele (III:2 and III:7) (incomplete penetrance). The minor allele frequency within the Maltese population was determined to be 0.012 (1.19%) with a population frequency 2.38%. According to the NCBI dbSNP database (http://www.ncbi.nlm.nih.gov/SNP/snp_ref.cgi?rs=11033026), and HapMap, this frequency compares with that found in Sub-Saharan Africans, African-Americans and Han Chinese from Beijing (minor allele frequencies 0.336, 0.115, 0.012, respectively), and was absent in European Caucasians. This suggests a founder effect in the Maltese population, complementing other previously reported studies on other human diseases (Farrugia et al., 2007; Koziell et al., 2002).

Since this variant was found in an exon but does not result in an amino acid change, we hypothesised that it could affect pre-mRNA splicing resulting in a different protein isoform.

To test this hypothesis at the transcriptional level an online Bioinformatics predictive tool was used to identify any possible exon splicing enhancers (ESEs) at this region as described by Cartegne et al (2002) (http://rulai.cshl.edu/tools/ESE2/index.html) . The G/A variant was found to abolish an ESE motif (TGAGGA > TGAAGA) for the SR protein (SRp55) with a score of 2.817 (threshold 2.676), in the presence of the A allele. Another online application RESCUE-ESE (http://genes.mit.edu/fas-ess/) did not predict any possible ESEs at this locus (Wang et al., 2004).
The experimental approach to test this hypothesis involved the use of an *in vitro* exon-trapping vector where the whole exon 9 and adjacent introns were inserted into a vector yielding a hybrid construct made up of two vector β-globin exons flanking CD44 exon 9 and adjacent introns. Following transfection into mammalian cells, the construct was transcribed under the control of a SV40 promoter and spliced. The mRNA derived from this construct was extracted and reverse transcribed followed by specific amplification using cDNA as template and specific primers to β-globin exons (SD6 and SA2). The spliced transcripts were analysed by agarose gel (Vidal et al., 2009). Our results showed that in the presence of the A allele only one transcript (261bp) was weakly amplified in both COS-7 and HeLa cells and was completely absent in RAW264.7 macrophages. DNA sequencing confirmed that this transcript did not contain any part of CD44 exon 9, and was entirely made up of vector exon sequences, suggesting skipping of exon 9. Two transcripts were amplified in the presence of the G allele (378bp and 261bp).

4. Conclusion

In this study, two polymorphisms with a population frequency of less than 5.0%were identified by linkage analysis in two extended Maltese families with a highly penetrant form of osteoporosis. *In vitro* functional studies confirmed that these polymorphisms might increase the individual's susceptibility to osteoporosis. This study adds to the existent knowledge of the complex pathophysiology involved in disorders such as osteoporosis. This knowledge is useful for the development of more targeted and individualised treatments. Our results added to the increasing evidence that rare but functional polymorphisms are also responsible for disorders such as osteoporosis, and also that using extended families with extreme phenotypes increases the chance to identify the responsible genes. Computer technology and the internet contribute significantly to the outcome of these studies. Both technologies were important tools for researchers throughout the whole study starting from planning and design of experiments, analysis of data and interpretation of results.

5. Acknowledgements

We would like to thank the families that participated in this study for their collaboration and Dr Raymond Galea for his support in patient recruitment. We also thank Dr Andrew Verner and the staff at the Genotyping Facility McGill University and Genome Quebec Innovation Centre, Montreal Canada for the STR genotyping and Dr Marisa Cassar at MLS BioDNA Ltd, Malta, for DNA sequencing of genes. We would like to thank Dr. Pierre Schembri Wismayer M.D., Ph.D. and Dr. Anthony Fenech Ph.D., B.Pharm. (Hons.), for the use of facilities at the Department of Anatomy and Cell Biology, and Department of Pharmacology and Clinical Therapeutics, University of Malta, and Prof Junko Oshima, Department of Pathology, University of Washington, Washington, USA, who generously donated the pSPL3 plasmid. This project was approved by the Research Ethics Committee, and supported by the Research Fund Committee of the University of Malta.

Glossary	Allele	Alternative states of genes only identical if their base sequences are identical
	Body Mass Index (BMI)	A statistic of the relationship between weight and height = body weight divided by height squared
	Bone mineral density (BMD)	A measure of bone density usually measured by x-ray techniques
	Haplotype	A set of variants (SNPs or STRs) that are inherited together as a single block on a linear chromosome
	Imprinting	Expression of genes depending upon the parent of origin
	Linkage disequilibrium (LD)	Groups of markers or genes on the same linear chromosome that are inherited together more often than expected by chance as long as genetic recombination does not take place between them. LD can be used to locate genes associated with phenotype
	Locus Heterogeneity	Variability of chromosomal regions involved between different subjects
	Penetrance	The percentage of individuals that express a trait determined by gene/s
	Phenocopy	A phenotyping change that mimics the expression of a mutation usually resulting from effects of the environment
	Segregate	Separation of homologous chromosomes at random during meiosis
	SNP	A difference in a single nucleotide at a particular DNA site
	STR	Short tandem repeat variations differing between different individuals in the number of repeated sequences eg: (CACACA) or (CACACACACA). Used as markers in forensics for identification.

6. References

Berger, M., Mattheisen, M., Kulle, B., Schmidt, H., Oldenberg, J., Bickeboller, H., Walter, U., Lindner, T.H., Strauch, K., & Schambeck, C.M. (2005). High factor VIII levels in venous thromboembolism show linkage to imprinted loci on chromosomes 5 and 11, Blood, Vol.105, No.2, (January 2005), pp. 638 - 644, ISSN 0006-4971

Borg, J., Papadopoulos, P., Georgitsi, M., Gutiérrez, L., Grech, G., Fanis, P., Phylactides, M., Verkerk, A.J., van der Spek, P.J., Scerri, C.A., Cassar, W., Galdies, R., van Ijcken, W., Ozgür, Z., Gillemans, N., Hou, J., Bugeja, M., Grosveld, F.G., von Lindern, M., Felice, A.E., Patrinos, G.P.. & Philipsen, S. (2010) Haploinsufficiency for the erythroid transcription factor KLF1 causes hereditary persistence of fetal hemoglobin. Nat Genet, Vol.42, No.9, (September 2010), pp. 801-805, ISSN 1061-4036

Brunkow, M.E., Gardner, J.C., Van Ness, J., Paeper, B.W., Kovacevich, B.R., Proll, S., Skonier, J.E., Zhao, L., Sabo, P.J., Fu, Y-H., Alisch, R.S., Gillett, L., Colbert, T., Tacconi, P., Galas, D., Hamersma, H., Beighton, P., & Mulligan, J.T. (2001) Bone dysplasia sclerosteosis results from loss of the SOST gene product, a novel cystine knot-containing protein. Am J Hum Genet, Vol.68, No.3, (March 2001), pp. 577-589, ISSN 0002-9297

Cartharius, K., Frech, K., Grote, K., Klocke, B., Haltmeier, M., Klingenhoff, A., Frisch, M., Bayerlein, M., & Werner, T. (2005) MatInspector and beyond: promoter analysis based on transcriptional factor binding sites. Bioinformatics, Vol.21, No.13, (July 2005), pp. 2933 - 2942, ISSN 1367-4803

Cartegni, L., Chew, S.L., & Krainer A.R. (2002) Listening to silence and understanding nonsense: exonic mutations that affect splicing. Nat Rev Genet, Vol.3, No.4, (April 2002), pp. 285-298, ISSN 1471-0056

Cassar, M. (2008) Linkage analysis in a familial case of idiopathic epilepsy and its implications in drug development. PhD Dissertation, University of Malta

Farrugia, R., Scerri, C.A., Attard Montalto, S., Parascandalo, R., Neville, B.G.R., & Felice, A.E. (2007) Molecular genetics of the tetrahydrobiopterin (BH4) deficiency in the Maltese population. Mol Genet Metab, Vol.90, No.3, (March 2007), pp. 277 - 283, ISSN 1096-7192

Gianfrancesco, F., Esposito, T., Ombra, M.N., Forabosco, P., Maninchedda, G., Fattorini, M., Casula, S., Vaccargiu, S., Casu, G., Cardia, F., Deiana, I., Melis, P., Falchi, M., & Pirastu, M. (2003) Identification of a novel gene and a common variant associated with uric acid nephrolithiasis in a Sardinian genetic isolate. Am J Hum Genet, Vol.72, No.6, (June 2003), pp. 1479 - 1491, ISSN 0002-9297

Gong Y, Slee, R.B, Fukai, N., Rawadi, G., Roman-Roman, S., Reginato, A.M., Wang, H., Cundy, T., Glorieux, F.H., Lev, D., Zacharin, M., Oexle, K., Marcelino, J., Suwairi, W., Heeger, S., Sabatakos, G., Apte, S., Adkins, W.N., Allgrove, J., Arslan-Kirchner, M., Batch, J.A., Beighton, P., Black, G.C., Boles, R.G., Boon, L.M., Borrone, C., Brunner, H.G., Carle, G.F., Dallapiccola, B., De Paepe, A., Floege, B., Halfhide, M.L., Hall, B., Hennekam, R.C., Hirose, T., Jans, A., Jüppner, H., Kim, C.A., Keppler-Noreuil, K., Kohlschuetter, A., LaCombe, D., Lambert, M., Lemyre, E., Letteboer, T., Peltonen, L., Ramesar, R.S., Romanengo, M., Somer, H., Steichen-Gersdorf, E., Steinmann, B., Sullivan, B., Superti-Furga, A., Swoboda, W., van den Boogaard, M.J., Van Hul, W., Vikkula, M., Votruba, M., Zabel, B., Garcia, T., Baron, R., Olsen, B.R., & Warman, M.L; Osteoporosis-Pseudoglioma Syndrome Collaborative Group. (2001) LDL receptor-related protein 5 (LRP5) affects bone accrual and eye development. Cell, Vol.107, No.7, (July 2001), pp. 513 - 523, ISSN 0092-8674

Hanson, R.L., Ehm, M.G., Pettitt, D.J., Prochazka, M., Thompson, D.B., Timberlake, B., Foroud, T., Kobes, S., Baier, L., Burns, D.K., Almasy, L., Blangero, J., Garvey, W.T., Bennett, P.H., & Knowler, W.C. (1998) An autosomal genomic scan for loci linked with type II diabetes mellitus and bone-mass index in Pima Indians. Am J Hum Genet, Vol.63, No.4, (October 1998), pp. 1130 - 1138, ISSN 0002-9297

Hayer, S., Steiner, G., Görtz, B., Reiter, E., Tohidast-Akrad, M., Amling, M., Hoffmann, O., Redlich, K., Zwerina, J., Skriner, K., Hilberg, F., Wagner, E.F., Smolen, J.S., & Schett, G. (2005) CD44 is a determinant of inflammatory bone loss. J Exp Med, Vol.201, No.6, (March 2005), pp. 903 - 914, ISSN 0040-8724

Heiman, G.A. (2005) Robustness of case-control studies to population stratification. Cancer Epidemiol Biomarkers Prev, Vol.14, No.6, (June 2005), pp. 1579 - 1582, ISSN 1055-9465

Kambouris, M. (2005) Target gene discovery in extended families with type 2 diabetes mellitus. Atheroscler Suppl, Vol.6, No.2, (May 2005), pp. 31 - 36, ISSN 1567-5688

Kania, J.R., Kehat-Stadler, T., & Kupfer, S.R. (1997) CD44 antibodies inhibit osteoclast formation. J Bone Miner Res. Vol.12, No.8, (August 1997), pp. 1155 - 1164, ISSN 0884-0431

Khan, A.A., Bachrach, L., Brown, J.P., Hanley, D.A., Josse, R.G., Kendler, D.L., Leib, E.S., Lentle, B.C., Leslie, W.D., Lewiecki, E.M., Miller, P.D., Nicholson, R.L., O'Brien, C., Olszynski, W.P., Theriault, M.Y., & Watts, N.B. (2004) Standards and guidelines for performing central dual-energy X-ray absorptiometry in premenopausal women, men, and children. J Clin Densitom, Vol.7, No.1, (Spring 2004), pp. 51 - 64, ISSN 1094-6950

Kong, A., & Cox, N.J. (1997) Allele sharing models: LOD scores and accurate linkage tests. Am J Hum Genet, Vol.61, No.5, (November 1997), pp. 1179 - 1188, ISSN 0002-9297

Koziell, A., Grech, V., Hussain, S., Lee, G., Lenkkeri, U., Tryggvason, K., & Scambler, P. (2002) Genotype/phenotype correlations of NPHS1 and NPHS2 mutations in nephrotic syndrome advocate a functional inter-relationship in glomerular filtration. Hum Mol Genet, Vol.11, No.4, (February 2002), pp. 379 - 388, ISSN 0964-6906

Kruglyak, L., Daly, M.J., Reeve-Daly, M.P., & Lander, E. (1996) Parametric and non-parametric linkage analysis, a unified multipoint approach. Am J Hum Genet, Vol.58, No.6, (June 1996), pp. 1347 - 1363, ISSN 0002-9297

Lander, E., & Kruglyak, L. (1995) Genetic dissection of complex traits, guidelines for interpreting and reporting linkage results. Nat Genet, Vol.11, No.3, (November 1995), pp. 241 - 247, ISSN 1061-4036

Lindner, T.H., & Hoffmann, K. (2005) EasyLINKAGE: a PERL script for easy and automated two-/multipoint linkage analyses. Bioinformatics, Vol.21, No.3, (February 2005), pp. 405 - 407, ISSN 1367-4803

Markianos, K., Daly, M.J., & Kruglyak, L. (2001) Efficient multipoint linkage analysis through reduction of inheritance space. Am J Hum Genet, Vol.68, No.4, (April 2001), pp. 963 - 977, ISSN 0002-9297

Nyholt, D.R. (2002) GENEHUNTER, Your 'one-stop shop' for statistical genetic analysis? Hum Hered, Vol.53, No.1, (March 2002), pp. 2 - 7, ISSN 0001-5652

O'Connell, J.R., & Weeks, D.E. (1998) PedCheck: a program for identification of genotype incompatibilities in linkage analysis. Am J Hum Genet, Vol.63, No.1, (July 1998), pp. 259 - 266, ISSN 0002-9297

Ott, J. (1991) Analysis of human genetic linkage. John Hopkins University Press, ISBN 0-801-842573, USA

Peltonen, L. (2000) Positional cloning of disease genes: Advantages of genetic isolates. Hum Hered, Vol.50, No.1, (January 2000), pp. 65 - 75, ISSN 0001-5652

Rozen, S., & Skaletsky, H.J. (2000) Primer3 on the WWW for general users and for biologist programmers. In: Bioinformatics Methods and Protocols: Methods in Molecular Biology, S. Krawetz, & S. Misener, (Ed.), 365-386, Humana Press, ISBN 0896037320,Totowa, NJ, USA

Seelow, D., Schwarz, J.M., & Schuelke, M. (2008) GeneDistiller--distilling candidate genes from linkage intervals. PLoS One. Vol.3, No.12, (December 2008), e3874, ESSN 1932-6203

Strauch, K., Fimmers, R., Baur, M.P., & Wienker, T.F. (2003) How to model a complex trait. Hum Hered, Vol.55, No.4, (October 2003), pp. 202 - 210, ISSN 0001-5652

Strauch, K., Furst, R., Ruschendorf, F., Windemuth, C., Dietter, J., Flaquer, A., Baur, M.P., & Wienker, T.F. (2005) Linkage analysis of alcohol dependence using MOD scores. BMC Genet, Vol.6, Suppl.1, (December 2005), S162, ESSN 1471-2156

Styrkarsdottir, U., Cazier, J.B., Kong, A., Rolfsson, O., Larsen, H., Bjarnadottir, E., Johannsdottir, V.D., Sigurdadottir, M.S., Bagger, Y., Christiansen, C., Reynisdottir, I., Grant, S.F.A., Jonasson, K., Frigge, M.L., Gulcher, J.R., Sigurdsson, G., & Stefansson, K. (2003) Linkage of osteoporosis to chromosome 20p12 and association to BMP2. PLoS Biol, Vol.1, No.3, (November 2003), pp. 351 - 360, ISSN 1544 - 9173

Thiele, H., & Nürnberg, P. (2004) HaploPainter: a tool for drawing pedigrees with complex haplotypes. Bioinformatics. Vol.21, No.8, (April 2004), pp. 1730-1732, ISSN 1367-4803

Vidal, C., Borg, J., Xuereb-Anastasi, A., & Scerri, C.A. (2009a) Variants within protectin (CD59) and CD44 genes linked to an inherited haplotype in a family with coeliac disease. Tissue Antigens, Vol.73, No.3, (March 2009), pp. 225 - 235, ISSN 0001-2815

Vidal, C., Cachia, A., & Xuereb-Anastasi, A. (2009) Effects of a synonymous variant in exon 9 of the CD44 gene on pre-mRNA splicing in a family with osteoporosis. Bone Vol.45, No.4, (October 2009), pp. 736 - 742, ISSN 8756-3282

Vidal, C., Galea, R., Brincat, M., & Xuereb-Anastasi, A. (2007) Linkage to chromosome 11p12 in two Maltese families with a highly penetrant form of osteoporosis. Eur J Hum Genet, Vol.15, No.3, (March 2007), pp. 800 - 809, ISSN 1018-4813

Wang, Z., Rolish, M. E., Yeo, G., Tung, V., Mawson, M. & Burge, C. B. (2004). Systematic identification and analysis of exonic splicing silencers. Cell Vol.119, No.6, (December 2004), pp. 831-845, ISSN 0092-8674

Wright, A., Charlesworth, B., Rudan, I., Carothers, A., & Campbell, H. (2003). A polygenic basis for late-onset disease. Trends Genet, Vol.19, No.2, (February 2003), pp. 97 - 106, ISSN 0168-9525

Wright, A.F., Carothers, A.D., & Pirastu, M. (1999) Population choice in mapping genes for complex diseases. Nat Genet, Vol.23, No.4, (December 1999), pp. 397 - 404, ISSN 1061-4036

Xu, J., & Meyers, D.A. (1998) Lod Score Analysis, In : Approaches to Gene Mapping in Complex Human Diseases, Haines, J.L., & Pericak-Vance, M.A., pp. 253 - 272, Wiley-Liss Inc., ISBN 0-471-17195-6, USA

Zlotogora, J. (2007) Multiple mutations responsible for frequent genetic diseases in isolated populations. Eur J Hum Genet, Vol.15, No.1, (January 2007), pp. 272 - 278, ISSN 1018-4813

3

Guide to Genome-Wide Bacterial Transcription Factor Binding Site Prediction Using OmpR as Model

Phu Vuong and Rajeev Misra
Arizona State University
USA

1. Introduction

Gene expression regulation in a cell plays a crucial role in the cellular response to environmental cues and other important biological processes (Bauer et al., 2010). A major mechanism of gene expression regulation is the binding of transcription factor (TF) protein to a specific DNA sequence in the regulatory region of a gene, thereby activating or inhibiting its transcription (Zhou & Liu, 2004). A TF often regulates multiple genes whose binding sites have similar but not identical sequences (Zhang et al., 2009). There is, however, a short, recurring pattern among the promoter sequences called a motif, and it is this motif that a TF recognizes and interacts with (D'haeseleer, 2006b). It is important to identify the set of genes a TF modulates, called its regulon, as this will advance our understanding of the regulatory network of an organism (D'haeseleer, 2006b; Tan et al., 2005). One way to identify the regulon is to determine a TF's motif and subsequently use the motif to search for other candidate genes regulated by the TF.

Traditionally, TF binding sites (TFBSs) are determined by various experimental approaches. Mutagenesis, DNase footprinting, gel-shift, and reporter construct assays are common methods for identifying the binding sites upstream of individual genes, but the throughput of these techniques is low (D'haeseleer, 2006b; Ladunga, 2010). In recent years, chromatin immunoprecipitation (ChIP) and systematic evolution of ligands by exponential enrichment (SELEX) are available to study protein-DNA interactions in a high throughput manner. Chromatin-immunoprecipitation of DNA cross-linked to a TF can be hybridized to a microarray (ChIP-chip) or sequenced (ChIP-seq) to obtain the TF's cognate binding sites on the whole genomic scale (Homann and Johnson, 2010; Ladunga, 2010; Stormo, 2010). SELEX is an *in vitro* technique that measures the binding affinities of TFs for synthetic, randomly generated oligonucleotides, usually 10-30 bp long (D'haeseleer, 2006b; Ladunga, 2010; Stormo, 2010). Sequences that strongly bind to a TF in question will be selectively amplified for later identification (Schug, 2008).

The major drawback of experimental approaches to determine TF recognition motifs is the time required and the relative high cost (Zhou & Liu, 2004). Moreover, some methods have specific requirements. For example, ChIP requires antibodies and certain growth conditions under which the transcription regulator is active (Tan et al., 2005). Even if a biologist can satisfy the requirements, the resolution of the regions containing the binding sites can span

from 30-50 bp (for SELEX) to a few hundreds bp (for ChIP-chip), making the extraction of the consensus motif from the collected sequences not an easy task (Stormo, 2010).
Thankfully, in recent times, a new, *in silico* approach to identify TF binding sites became available. Many bioinformatics programs—the number ranges from 120 (Wei & Yu, 2007) to over 200 (Ladunga, 2010)—have been created to help biologists predict DNA binding motifs from the enormous amounts of sequence and gene expression data generated from advances in high-throughput genomic sequencing and gene expression analysis techniques.

1.1 Pattern matching and pattern discovery

There are two types of motif searches and the type dictates which programs one uses. In the first type, known as unsupervised motif finding or *de novo*, *ab initio*, or pattern discovery, a researcher wants to know the consensus pattern in a set of orthologous genes, genes in a common pathway, or transcriptionally co-regulated genes or operons from an experiment (Mrazek, 2009). The genes presumably share some binding sequence for a common TF and the task is to discover the conserved, statistically over-represented motif in the regulatory regions (Mrazek, 2009). In the second type, known as supervised motif finding or pattern matching, the DNA binding motif for a TF has been determined—either predicted *de novo* or experimentally identified—and the goal is to find which other genes in the genome have a similar motif in their promoter (Mrazek, 2009).
Because pattern discovery and pattern matching are fundamentally different tasks, there are two classes of motif prediction programs, each implementing different algorithms to solve their respective problem. For *de novo* motif discovery programs, the goal is to iteratively find a set of 12-20-bp sequence motifs that are most significantly similar to each other (Mrazek, 2009), usually with an enumeration, expectation maximization, or Gibbs sampling algorithm. Representative programs of this class include AlignACE, MEME, BioProspector, MDScan and MotifSampler (Hu et al., 2005). This is the extent of our coverage on pattern discovery in this chapter. For more information, see Ladunga, 2010; MacIsaac & Fraenkel, 2006; Mrazek, 2009; Stormo, 2000; and Wei & Yu, 2007. For more details on the algorithms, see Das & Dai, 2007; D'haeseleer, 2006a; Pavesi et al., 2004; and Stormo, 2010.

1.2 What's covered

The rest of this chapter discusses pattern matching with a focus on prokaryotes. Eukaryotes are not covered because transcription regulation is substantially different between these two groups (Quest et al., 2008). Promoters of prokaryotes are typically less than 500 bp and are more likely to be palindromic, whereas those in eukaryotes can extend tens of thousands of nucleotides (Thompson et al., 2007). Another difference is that prokaryotic TFBSs are a few hundred bp upstream of translational start site and can overlap or appear in tandem, whereas in eukaryotes, they can be kilobases away (Bulyk, 2003; Yanover et al., 2009). Lastly, in prokaryotes, gene regulation occurs mainly at the transcriptional level (Yanover et al., 2009). In eukaryotes, multiple TFs coordinately bind to relatively short binding sites in the promoter of a single gene to regulate its expression (Thompson et al., 2007; Yoshida et al., 2006; Zaslavsky & Singh, 2006). Also in eukaryotes, the genome is bigger with more non-coding sequences and the regulatory elements can be located upstream of the gene, within it, or even downstream of it (Bulyk, 2003). With eukaryote gene regulation being more complex, motif finding programs work significantly better on lower organisms than on higher organisms (Das & Dai, 2007).

The chapter is intended to be a pragmatic guide for microbiologists. As such, it does not cover algorithms in details and technical mathematical formulas. Instead, it presents a high-level conceptual overview of the key concepts researchers need to know in order to effectively use the available bioinformatics tools to locate TF binding sites in sequenced prokaryote genomes. Online databases of prokaryote gene expression regulation information are introduced next, followed by pattern matching programs designed for or tested on prokaryotes. Finally, the chapter concludes by offering practical strategies and tips to improve the specificity and sensitivity of the results.

Along the way, the global transcriptional regulator OmpR in *Escherichia coli* will be used in examples throughout the chapter. OmpR is a cytosolic response regulator, and together with the membrane-bound histidine sensor kinase EnvZ, constitute a prototypical two-component signal transduction system in bacteria. Our lab is currently using bioinformatics to identify novel target genes of OmpR in the *E. coli* genome.

2. Motif representation

As mentioned before, a TF binds to different DNA sequence variations to modulate the expression of their target genes. This degeneracy of the binding sequences allows different levels of gene regulation to be achieved (D'haeseleer, 2006b). For instance, OmpR's DNA binding properties fluctuate with the extent of covalent modifications, leading to changes in the DNA binding affinity and/or its DNA binding "signature", and thus broadening its motif definition. In *E. coli* one of the genes regulated by OmpR, *ompF*, illustrates the transcriptional regulator's broad recognition signature. *In vivo* and *in vitro* experiments have shown that two OmpR molecules bind to each of the four sites in the promoter of the *ompF* gene in a tandem manner (Harlocker et al., 1995; Yoshida et al., 2006):

F1:	TTTACTTTTGGTTACATATT
F2:	TTTTCTTTTTGAAACCAAAT
F3:	TTATCTTTGTAGCACTTTCA
F4:	GTTACGGAATATTACATTGC

Many pattern matching programs take a set of TFBS sequences, such as the OmpR binding sequences above, as input and internally convert it to a matrix representation for genome scanning. A few programs, such as MAST, require the matrix directly, which can be constructed using one of the matrix utility programs discussed later.

Conceptually, a motif matrix is a table of 4 rows by *n* columns, where *n* is the length of the TFBS sequences, that tabulates the frequency information of the nucleotides at each position. The four rows correspond to the four nucleotides A, T, G, C. Each column in the table holds the occurrence frequency of each base at that motif position. Bases that occur more frequently at a position/column have a higher number. See Fig. 1(a) for the matrix representation of the four F1-F4 OmpR binding sequences shown above.

The matrix in Fig. 1(a) is called a position frequency matrix. In actual practice, bioinformatics programs add values like pseudocounts (to avoid zero, which is undefined for some mathematical functions used in the algorithm) and background model probabilities (to account for genome differences like GC content) to each frequency number

(Mrazek, 2009). It is this more sophisticated matrix, called a position-specific score matrix (PSSM) or position weight matrix (PWM), that is actually used by pattern matching programs during genome scanning. See Fig. 1(b).

Matrices are generally used to represent more degenerate (that is, less conserved) TFBS sequences (Mrazek, 2009). When the consensus pattern is more conserved, one may model the motif using the International Union of Pure and Applied Chemistry (IUPAC) codes (D'haeseleer, 2006a). The IUPAC system defines 11 new single-letter codes that represent more than one nucleotide (see Table 1). For example, the *ompF* binding sites could be neatly represented using the IUPAC alphabet as KTWWCKKWDKRDHACHWWNH [see Fig. 1(c)]. The *K* is a "wild card" code for guanine or thymine; *W*, adenine or thymine; and so on.

IUPAC code	Matches Nucleotide(s)
A	A
C	C
G	G
T	T
R	A or G
Y	C or T
W	A or T
S	G or C
M	A or C
K	G or T
H	A, C or T
B	C, G or T
V	A, C or G
D	A, G or T
N	A, C, G or T
. or -	(gap)

Table 1. IUPAC codes for describing more conserved transcription factor binding consensus sequences (Pavesi et al., 2004).

Another way to represent more conserved motifs is via regular expression, or regex, a complex but highly flexible language for describing text patterns in the computer field (Mrazek, 2009). One simple regular expression that describes the four OmpR-binding sites upstream of *ompF* is: [TG]T[AT][AT]C[TG][TG][AT][ATG][TG][GA][ATG][ATC]AC [ATC][AT][AT][ATGC][ATC]

Each pair of brackets specifies one nucleotide and the bases in the brackets specify the allowable nucleotides. There are slightly different flavors of the regex language that implement slightly different features, so be sure to check the documentation accompanying a pattern matching program to find out the features supported.

Note that both IUPAC codes and regular expression allow multiple bases to be specified at a nucleotide position, but all the valid bases are assumed to occur with the same frequency. Because the set of DNA sequences recognized by a TF is often degenerate and nucleotide frequency information is helpful in pattern matching, matrices are more often used and are supported by many programs.

F1	T	T	T	A	C	T	T	T	T	G	G	T	T	A	C	A	T	A	T	T
F2	T	T	T	T	C	T	T	T	T	T	G	A	A	A	C	C	A	A	A	T
F3	T	T	A	T	C	T	T	T	G	T	A	G	C	A	C	T	T	T	C	A
F4	G	T	T	A	C	G	G	A	A	T	A	T	T	A	C	A	T	T	G	C
(a) A	0.00	0.00	0.25	0.50	0.00	0.00	0.00	0.25	0.25	0.00	0.50	0.25	0.25	1.00	0.00	0.50	0.25	0.50	0.25	0.25
T	0.75	1.00	0.75	0.50	0.00	0.75	0.75	0.75	0.50	0.75	0.00	0.50	0.50	0.00	0.00	0.25	0.75	0.50	0.25	0.50
C	0.00	0.00	0.00	0.00	1.00	0.00	0.00	0.00	0.00	0.00	0.00	0.00	0.25	0.00	1.00	0.25	0.00	0.00	0.25	0.25
G	0.25	0.00	0.00	0.00	0.00	0.25	0.25	0.00	0.25	0.25	0.50	0.25	0.00	0.00	0.00	0.00	0.00	0.00	0.25	0.00
(b) A	-8.65	-8.65	-0.54	0.46	-8.65	-8.65	-8.65	-0.54	-0.54	-8.65	0.46	-0.54	-0.54	1.46	-8.65	0.46	-0.54	0.46	-0.54	-0.54
T	1.04	1.46	1.04	0.46	-8.65	1.04	1.04	1.04	0.46	1.04	-8.65	0.46	0.46	-8.65	-8.65	-0.54	1.04	0.46	-0.54	0.46
C	-8.65	-8.65	-8.65	-8.65	2.87	-8.65	-8.65	-8.65	-8.65	-8.65	-8.65	-8.65	0.87	-8.65	2.87	0.87	-8.65	-8.65	0.87	0.87
G	0.87	-8.65	-8.65	-8.65	-8.65	0.87	0.87	-8.65	0.87	0.87	1.87	0.87	-8.65	-8.65	-8.65	-8.65	-8.65	-8.65	0.87	-8.65
(c) IUPAC	K	T	W	W	C	K	K	W	D	K	R	D	H	A	C	H	W	W	N	H

Fig. 1. The first four rows, labeled F1-F4, contain the four sites upstream of the *ompF* gene where OmpR binds in *Escherichia coli*. (a) The position frequency matrix representation of the same *ompF* sequences. Each column contains the frequencies of occurrence of the nucleotides in each corresponding F1-F4 sequence position. (b) The position weight matrix representation of the *ompF* F1-F4 sequences. The weight matrix is derived from the frequency matrix in (a). The values are calculated by taking the log of the frequency values divided by background model values. It is this position weight matrix that is actually used by pattern matching programs during execution. (c) The consensus motif of the four F1-F4 sequences in IUPAC codes.

3. How pattern matching programs work

Whether a motif is given as a regular expression, an IUPAC consensus sequence, or a matrix, a pattern matching program looks for the motif by scanning a genome on both the sense and antisense strands from the 5′ to 3′ end (MacIsaac & Fraenkel, 2006). See Fig. 2. Typically, the default is to check only the intergenic regions; coding regions are skipped over. A window with a width equal to the length of the motif slides over the genome one base at a time. At each iteration, the sequence in the window is checked against the given motif for a match.

For regular expression or IUPAC motif, each nucleotide in the window is checked to see if that nucleotide is allowed at that position. If the number of mismatches is at or below a certain limit, the sequence is considered a match and returned. If the motif is given as a matrix, the sequence is scored against the matrix. The score for that sequence is calculated by summing the weight score at each position. If the score is at or above a certain threshold, that sequence is considered similar to the motif and a match is found. The score measures how closely the candidate sequence matches the motif modeled by the position weight matrix and how likely the candidate happens to be a random genomic background sequence.

4. Motif databases and utilities

To use motif matching programs to discover candidate genes modulated by a TF, the TF's motif is required. One can look in the literature to compile a list of the reported binding site sequences, or better yet, one can search online databases of sequenced genomes and gene regulation information, usually curated from primary journals. There are general databases covering the prokaryotes and specialized ones for particular bacterial strains (see Table 2). For instance, PRODORIC contains close to 3,000 TFBSs and over 2,000 genes for multiple bacteria species (Grote et al., 2009). Another resource containing information on transcription factors and their target genes, but for *Escherichia coli* K-12 only, is RegulonDB (Gama-Castro et al., 2011).

... CTTTTTGAAAATACGCAACGGCCATTTTTGCACTTAGATACAGATTTTCTGCGCTGTATTGCATTGATTTGAT...

a)	T	T	T	G	C	A	C	T	T	A	G	A	T	A	C	A	G	A	T	T
	K	T	W	W	C	K	K	W	D	K	R	D	H	A	C	H	W	W	N	H
	Y	Y	Y	N	Y	N	N	Y	Y	N	Y	Y	Y	Y	Y	Y	N	Y	Y	Y

b)	T	T	T	G	C	A	C	T	T	A	G	A	T	A	C	A	G	A	T	T
A	-8.65	-8.65	-0.54	0.46	-8.65	-8.65	-8.65	-0.54	-0.54	-8.65	0.46	-0.54	-0.54	1.46	-8.65	0.46	-0.54	0.46	-0.54	-0.54
T	1.04	1.46	1.04	0.46	-8.65	1.04	1.04	1.04	0.46	1.04	-8.65	0.46	0.46	-8.65	-8.65	-0.54	1.04	0.46	-0.54	0.46
C	-8.65	-8.65	-8.65	-8.65	2.87	-8.65	-8.65	-8.65	-8.65	-8.65	-8.65	-8.65	0.87	-8.65	2.87	0.87	-8.65	-8.65	0.87	0.87
G	0.87	-8.65	-8.65	-8.65	-8.65	0.87	0.87	-8.65	0.87	0.87	1.87	0.87	-8.65	-8.65	-8.65	-8.65	-8.65	-8.65	0.87	-8.65

Score: 1.04+1.46+1.04-8.65+2.87-8.65-8.65+1.04+0.46-8.65+1.87-0.54+0.46+1.46+2.87+0.46-8.65+0.46-0.54+0.46 = -28.38

Fig. 2. Overview of how pattern matching programs work given a motif represented using (a) IUPAC codes or (b) a position weight matrix. These programs slide an n-bp window (shaded and underlined), where n is the length of the motif ($n = 20$ in this example), over every single base in the genome. For each iteration, the sequence inside the window is (a) compared against the allowable nucleotides specified by the IUPAC motif, and if the number of mismatches is at or below a certain limit, the sequence is considered a match. For the matrix in (b), the score of an individual base in each column is looked up and summed, and if the total is at or over a certain threshold, the sequence is considered a match. The IUPAC consensus motif and the matrix are the same as those in Fig. 1(c) and (b). In (a), Y = a match, N = mismatch.

Database	Organism	Web Address	Reference
DBTBS	*Bacillus subtilis*	http://dbtbs.hgc.jp	Sierro et al., 2008
DPInteract	*Escherichia coli*	http://arep.med.harvard.edu/dpinteract	Robison et al., 1998
RegulonDB	*Escherichia coli*	http://regulondb.ccg.unam.mx	Gama-Castro et al., 2011
CoryneRegNet	*Corynebacterium*	http://www.CoryneRegNet.de	Baumbach, 2007
PRODORIC	*Prokaryotes*	http://www.prodoric.de	Grote et al., 2009
RegPrecise	*Prokaryotes*	http://regprecise.lbl.gov/RegPrecise	Novichkov et al., 2010a
RegTransBase	*Prokaryotes*	http://regtransbase.lbl.gov	Kazakov et al., 2007
KEGG	*Prokaryotes*	http://www.genome.ad.jp/kegg	Kanehisa et al., 2004

Table 2. Select databases of curated and annotated transcription factor binding sites and other gene expression regulation information in prokaryotes. Note that the KEGG database also covers eukaryotes.

Once a set of binding site sequences has been gathered, online tools are available to analyze and display the motif in those sequences (Table 3). D-MATRIX (Sen et al., 2009) is a web application that constructs alignment, frequency, and weight matrices and displays them.

The generated matrices can be exported for use as input into pattern matching programs. The site can also generate the regular expression and IUPAC representations of the consensus motif. WebLogo displays a motif graphically so that sequence similarity can easily be visualized (Crooks et al., 2004).

It is important that the gathered TF binding sites are high quality since inaccuracies will produce a subpar matrix and consequently, poor motif matching performance (Medina-Rivera et al., 2010; Wittkop et al., 2010). Inaccuracies in TFBS information could stem from the imprecise nature of experimental approaches since gel shift, DNase footprinting, ChIP-chip, and ChIP-seq do not precisely identify binding sequences (Wittkop et al., 2010).

Two programs aim to analyze and optimize binding sequences. The utility 'matrix-quality' quantifies the ability of a matrix to distinguish background sequences and find functional binding sites in a genome (Medina-Rivera et al., 2010). It works by combining theoretical and empirical score distributions (Medina-Rivera et al., 2010). Another utility, MoRAine, goes one step further by shifting nucleotides around and takes the reverse complement of each TFBS sequence to try to improve the matrix (Wittkop et al., 2010) .

Program	Platform	Web Address	Reference
D-MATRIX	Web	http://203.190.147.116/dmatrix	Sen et al., 2009
matrix-quality	Web; Unix	http://rsat.ulb.ac.be/rsat	Medina-Rivera et al., 2010
MoRAine	Web; Java	http://moraine.cebitec.uni-bielefeld.de	Wittkop et al., 2010
WebLogo	Web; Unix	http://weblogo.berkeley.edu	Crooks et al., 2004

Table 3. Utility programs to manipulate transcription factor binding site sequences: construct and display frequency and weight matrices, generate regular expressions and IUPAC consensus patterns, and check and improve alignment quality. All the programs run inside a web browser (Platform = Web). Some of them can also be downloaded and executed locally on the user's computer running Unix or Unix-like operating system (Platform = Unix) or locally inside a Java virtual machine (Platform = Java). The Web Address column shows where the programs can be located or downloaded.

5. Pattern matching programs

Once a list of high quality TF binding sites is in hand, it can be fed into the pattern matching programs listed in Table 4 to find novel binding sites. All the listed programs are designed for prokaryotes or they have been tested with bacteria. This section briefly describes each motif matching program.

STAMP (Mahony & Benos, 2007) is not a true pattern matching program in that it does not scan genomes. Instead, it finds regulatory sequences deposited in motif databases that are most similar to a user-supplied set of binding sequences. It also performs multiple alignments on the supplied binding motifs and builds trees of the evolution of TF binding motifs.

MAST (Motif Alignment & Search Tool) (Bailey & Gribskov, 1998), a component of the MEME Suite, is one of the early programs that perform pattern matching on nucleotide (and protein) sequences. The user can select one of the available genomes or upload a file containing up to one million nucleotides to search. The program requires an input file describing the matrix of the motif to search for.

Program	Platform	Web Address	Reference
CRoSSeD	Web	http://ibiza.biw.kuleuven.be/crossed/webtool.html	Meysman et al., 2011
EMBOSS > dreg	Web; Unix	http://emboss.sourceforge.net	Rice et al., 2000
dscan	Web	http://bayesweb.wadsworth.org/cgi-bin/dscan.pl	Thompson et al., 2005
FITBAR	Web	http://archaea.u-psud.fr/fitbar	Oberto, 2010
iMotifs	Mac	http://wiki.github.com/mz2/imotifs	Piipari et al., 2010
MAST	Web; Unix	http://meme.sdsc.edu/meme/mast-intro.html	Bailey & Gribskov, 1998
Motif Locator	Web	http://www.cmbl.uga.edu/software.html	Mrazek et al., 2008
MyPattern Finder	Web	http://www.nii.ac.in/~deepak/RegAnalyst	Sharma et al., 2009
PatScan	Unix	http://ftp.mcs.anl.gov/pub/Genomics/PatScan	Dsouza et al., 1997
Pattern Locator	Web; Unix	http://www.cmbl.uga.edu/software.html	Mrazek & Xie, 2006
PhyloScan	Web	http://bayesweb.wadsworth.org/phyloscan	Palumbo & Newberg, 2010
PredictRegulon	Web	http://www.cdfd.org.in/predictregulon	Yellaboina et al., 2004
RegPredict	Web	http://regpredict.lbl.gov	Novichkov et al., 2010b
RSAT	Web	http://rsat.ulb.ac.be/rsat	Thomas-Chollier et al., 2008
SITECON	Web	http://wwwmgs.bionet.nsc.ru/mgs/programs/sitecon	Oshchepkov et al., 2004
STAMP	Web	http://www.benoslab.pitt.edu/stamp	Mahony & Benos, 2007
Virtual Footprint	Web	http://www.prodoric.de/vfp	Munch et al., 2005

Table 4. Programs that can scan prokaryote genomes for transcription factor binding sites. All these programs can run over the web inside a browser (Platform = Web) except iMotifs, a MacOS X only application, and PatScan, a program that must be downloaded and run locally on Unix or Unix-like systems. Three of the web applications—dreg, MAST, and Pattern Locator—can also be downloaded and execute on Unix or Unix-like systems. The Web Address column shows where a program can be run or downloaded.

PatScan (Dsouza et al., 1997) is another early motif matching program. Even though it is designed to search protein sequences for motifs and nucleotide sequences for hairpins, pseudoknots, repeats, and other secondary structures, it could be used to search genomic DNA for TFBSs. Mismatches, insertions, and deletions are allowed. It runs on Unix systems only. A web version seems to be no longer available.

The program 'dreg' searches one or more sequences for a given motif described by a regular expression (Rice et al., 2000). It is one of the hundreds of tools comprising EMBOSS (European Molecular Biology Open Software Suite). EMBOSS' mission is to provide a place to bring together the rapid increase in the number of complete genomes and new sequence analysis software and make them publicly available as a suite. The tools perform sequence alignment, database searching with sequence patterns, nucleotide sequence patterns (CpG islands, repeats, etc.), and more.

To address the usability issue associated with PatScan and dreg, Pattern Locator was created (Mrazek & Xie, 2006). Its purpose is to find short sequence patterns in complete genomes. The input string uses a special syntax or it can be specified using the IUPAC alphabet. The flexible syntax allows the following to be specified: direct and inverted repeats, maximum number of mismatches allowed, direct or complementary DNA strand to search, and gaps.

Many other programs do not require motifs to be supplied in a special syntax. Motif Locator (Mrazek et al., 2008) takes a set of binding sequences, turns it into a matrix, and uses the matrix to search a genome for instances of the motif. MyPatternFinder (Sharma et al., 2009) finds exact or approximate occurrences of a motif from a selection of over 600 complete genomes using an exact search method and an alignment technique. Insertions and deletions are allowed. The program 'dscan' (Thompson et al., 2005) scans genome databases for statistically significant sites similar to the given motif. Two databases of *E. coli* and *Rhodopseudomonas palustris* intergenic regions are provided.

PredictRegulon (Yellaboina et al., 2004) is a web application that scans a prokaryote genome for potential target genes of a TF. The user picks a bacterial genome from a list of over 110 and supplies a set of aligned binding site sequences for the transcription factor. The program then scans the upstream sequences of all the genes in the selected genome, calculates a score for a potential binding site in each promoter, and outputs the site if the score is above the threshold cutoff value, which is taken to be the lowest score in the input sequence set. The output includes the binding site sequence, the name and description of the gene, and operon context and detailed information on the gene.

FITBAR (Fast Investigation Tool for Bacterial and Archaeal Regulons) (Oberto, 2010) is a matrix search program that scans whole Bacteria and Archaea genomes retrieved from the National Center for Biotechnology Information repository to discover sets of genes regulated by TFs. Unlike most other genome matching programs such as PredictRegulon, FITBAR does not find matches by arbitrary score cutoff values. It uses the log-odds and entropy-weighted search algorithms and Compound Importance Sampling (CIS) and Local Markov Method (LMM) to calculate the statistical significance of the predicted motifs (p-values). Aligned TFBS sequences can be supplied, or one of the 200 known prokaryotic matrices can be selected. Results are listed, along with a graphical depiction of the motif sequence location and the surrounding genes.

Like EMBOSS, RSAT (Regulatory Sequence Analysis Tools) (Thomas-Chollier et al., 2008) contains a collection of tools to analyze *cis*-acting regulatory elements in the noncoding sequences from over 600 genomes. The tools perform both pattern matching (and pattern discovery) and return information on individual genes, such as orthologs and DNA sequences. Namely, the five pattern matching programs—*dna-pattern*, *genome-scale-dna-*

pattern, *matrix-scan*, *patser*, and *genome-scale-patser*—allow one to search entire genomes or a set of sequences for occurrences of a motif represented as a regular expression, an IUPAC string, or a position weight matrix. Various statistical background models are available to allow the significance of the matches to be evaluated.

Virtual Footprint (Munch et al., 2005) is an online interactive environment to search and analyze TFBS, gapped or ungapped, in bacterial genomes. The TFBS pattern to search can be picked from a list of pre-defined matrix motifs, or a set of sequences, a regular expression, or IUPAC codes can be supplied. A match is assigned to a gene if possible and the genomic context is provided. The program can check if a match also occurs in the regulatory region of orthologous genes.

Like Virtual Footprint, iMotifs (Piipari et al., 2010) provides an integrated environment to visualize, analyze, and annotate sequence motifs. However, it does not run over the web. It is a Java-based application that runs on MacOS X only.

More advanced pattern matching programs incorporate the use of cross-species conservations during their genome search to enrich the predicted sites. Comparative genomics approach is predicated on the idea that TFs from related organisms regulate genes that tend to be conserved (Novichkov et al., 2010b). Presence of similar TFBSs upstream of orthologous genes increases the probability that the sites are functional binding sites (Novichkov et al., 2010b).

PhyloScan (Carmack et al., 2007; Palumbo & Newberg, 2010) is a web program that screens candidate sequences by using (1) aligned or unaligned sequence data from multiple species, even evolutionarily distant ones, (2) multiple sites within an intergenic region, and (3) q-values to predict more functional TFBSs, even weak ones, in a genome. The use of q-values is in contrast to conventional motif matching programs, which either score a candidate binding site against a training set of TFBS or evaluate the statistical significance of the candidate binding site using p-value.

Another program that takes the comparative genomics approach is RegPredict (Novichkov et al., 2010b), a web site that provides a visual environment for the discovery of genes regulated by a TF in prokaryotes. The site contains a large collection of known TFBS motifs gathered from the RegPrecise, RegTransBase, and RegulonDB databases and genomic sequences of major taxonomic groups of Bacteria. Any of the motifs can be selected, or the user can upload a set of aligned binding site sequences, and RegPredict will scan for the motif in up to 15 genomes simultaneously. (If the regulatory motif is not known, RegPredict can predict one *de novo* from user-supplied coregulated genes.) Candidate genes are grouped into different clusters based on the degree of conservation of regulatory interactions and then presented in a multi-pane user interface, along with the genomic context and gene function information, for the user to analyze.

Other advanced motif matching programs use DNA structure information to increase their performance. A factor that contributes to the specificity of the interaction between a TF and its binding site is the local conformation of the DNA site (Oshchepkov et al., 2004). Even though a TF often regulates multiple genes and the binding sites in the promoters of these genes show variations, certain conformational and physicochemical properties are conserved among these sites so that the TF can recognize the sites (Oshchepkov et al., 2004). Thus, these context-dependent TFBS properties can be used to improve the predictions of genes controlled by a TF (Oshchepkov et al., 2004).

SITECON (Oshchepkov et al., 2004) is a web application that can analyze and report 38 properties—major groove depth, bend, entropy change, to name a few—in a given set of DNA binding site sequences, and optionally, find binding sites in one or more DNA sequences using those properties.

CRoSSeD (Conditional Random fields of Smoothed Structural Data) (Meysman et al., 2011) is another program that leverages structure information. Specifically, it uses 12 structural scales, such as protein-induced deformability and stabilization energy, that are presumably relevant to binding site recognition in prokaryotes. (Scales are experimentally determined models for approximating regional DNA structure based on di- or trinucleotides.) Some of the novel binding sites found by CRoSSeD had low sequence similarity. A check with the literature and database indicated that they may be true binding sites. This shows that searching for binding sites based on structure information is a viable approach since these binding sites, with their weak motif, may be missed by traditional pattern matching programs.

6. EnvZ/OmpR regulon prediction

Our lab is currently using genetic, biochemical, and bioinformatics approaches to determine the set of genes regulated by OmpR in *E. coli*. A microarray experiment showed that the expression levels of 125 genes were significantly affected in an EnvZ-null background (Oshima et al., 2002). To help identify the genes that are directly modulated by OmpR, we searched the RegulonDB databank and found 23 OmpR binding sites for 11 genes, as listed in Table 5.

Using all of the OmpR binding sequences except *ecnB*'s (since it is not 20-bp long) as input, the pattern matching program Motif Locator detected 12,314 matches in the intergenic regions of the *E. coli* K12 genome. Since *E. coli* has over 4,200 genes (Blattner et al., 1997), the results clearly contained many false positives.

Gene	OmpR Binding Site	Gene	OmpR Binding Site
bolA	AACCTAAATATTTGTTGTTA	*micF*	CGAATATGATACTAAAACTT
nmpC	AACTTACATCTTGAAATAAT	*micF*	TTAAGATGTTTCATTTATCG
ompF	TTTACTTTTGGTTACATATT	*micF*	TATAGATGTTTCAAAATGTA
ompF	TTTTCTTTTTGAAACCAAAT	*ompC*	TTTACATTTTGAAACATCTA
ompF	CTTTATCTTTGTAGCACTTT	*ompC*	AGCGATAAATGAAACATCTT
ompF	GTTACGGAATATTACATTGC	*ompC*	AAAAGTTTTAGTATCATATT
csgD	TACATTTAGTTACATGTTTA	*fadL*	GAGCCAGAAAACCCTGTTTA
tppB	GTAACAGATTATTACAAAGG	*fadL*	TTAGATCATATTTGAAAAAA
flhD	AAAAATCTTAGATAAGTGTA	*fadL*	ACGTAACATAGTTTGTATAA
flhD	GGGCATTATCTGAACATAAA	*fadL*	AAATCACACTTAAAAATGAT
omrA	CACACCTCGTTGCATTTCCC	*ecnB*	AACATAAATAACAT
omrB	AACCTTTGGTTACACTTTGC		

Table 5. List of known OmpR binding sites and the corresponding genes. The list was compiled using RegulonDB.

To increase the specificity and reduce the number of matches returned, we picked 10 binding sites from five genes: *ompF*, *ompC*, *tppB*, *csgD*, and *fadL*. See Table 6. These sequences were chosen because each contains two direct repeats of the consensus motif GTTACANNNNN, which is derived from extensive studies on interactions between OmpR and *ompF* and *ompC* promoters (Harlocker et al., 1995; Yoshida et al., 2006). Note that we adjusted the alignment of the *csgD* and *fadL* binding sequences to better fit the consensus model. The *csgD* and *fadL* sequences from RegulonDB shown in Table 5 span from -57 to -38 and from +58 to +77 relative to the transcriptional start site, whereas the adjusted ones span from -59 to -40 and from 50 to 69, respectively.

Pattern matching analysis of the 10 sequences listed in Table 6 using Motif Locator found 110 matches, five of which were among the 125 genes affected in the microarray experiment: *ompF, flgL, ompC, rpoE,* and *cysC*. The same set of 10 sequences was fed into another pattern matching program, Virtual Footprint, which predicted 32 genes modulated by OmpR. Four genes were the same as those identified in the microarray data: *ompF, ydgR, ompC,* and *ygjU*. Only two genes were found by both Motif Locator and Virtual Footprint, *ompF* and *ompC*, showing that different programs return different results.

Gene	OmpR Binding Site
ompF	TTTACTTTTGGTTACATATT
ompF	TTTTCTTTTTGAAACCAAAT
ompF	CTTTATCTTTGTAGCACTTT
ompF	GTTACGGAATATTACATTGC
ompC	TTTACATTTTGAAACATCTA
ompC	AGCGATAAATGAAACATCTT
ompC	AAAAGTTTTAGTATCATATT
tppB	GTAACAGATTATTACAAAGG
csgD	GTTACATTTAGTTACATGTT
fadL	GTTACAGCACGTAACATAGT

Table 6. OmpR binding sequences that contain direct repeats of the GTTACANNNNN consensus motif, where N denotes any nucleotide.

The degenerate OmpR binding motif makes identification of new regulon member difficult. When the set of 10 sequences in Table 6 was used as input, Motif Locator predicted only half of the 12 known OmpR regulated genes: *bolA, ompF, csgD, micF, ompC,* and *fadL*, whereas Virtual Footprint returned four: *ompF, micF, ompC,* and *fadL*. This observation suggests that the run was too specific and more novel genes remain to be discovered. To find them, one can try different sets of input sequences, run other pattern matching programs, or make use of comparative genomics or published OmpR crystal structures (Kondo et al., 1997; Martínez-Hackert & Stock, 1997).

7. Conclusion

Like our own experience of using bioinformatics tools to study the OmpR regulon illustrates, comparative studies on the performance of motif discovery and matching programs found no single program works well on all data sets (MacIsaac & Fraenkel, 2006). In particular, a benchmark of four motif matching programs—RSA Tools, PRODORIC Virtual Footprint, RegPredict, and FITBAR—for their ability to discover potential binding sites for the transcriptional regulator NagC involved in N-acetylglucosamine metabolism in the *Escherichia coli* K12 MG1655 genome found that some tools uncover sites that others have missed (Oberto, 2010). Therefore, it is recommended that multiple tools be used instead of just one and the output from multiple programs be combined and compared in order to improve accuracy and gain confidence in the results (Das & Dai, 2007; Mrazek, 2009).

The easiest way to run the pattern matching programs—and other bioinformatics tools—is over the web inside a browser. However, in order to help keep the load on the web servers hosting these programs to a low level, some sites put a limit on the complexity of the jobs submitted. If a web site places such restriction, a desktop version of the program is usually provided for users to download and install or compile on their local computer. Many of the desktop programs run in a Java environment or on Unix or Unix-like system, such as Linux. Some Unix programs can run on Windows if the Linux-like environment Cygwin is set up first. However, it should be noted that setting up the required runtime environment and installing or compiling these programs take considerable effort and computer expertise. Also be aware that some desktop programs, especially those that run on Unix, are run from the command line; there is no graphical user interface.

The identification of a TF's binding motif and the identification of new target genes are difficult to do experimentally and computationally (Pavesi et al., 2004) because we do not completely understand the biology of gene regulation (Das & Dai, 2007). But it is hoped that the information in this chapter will make the task of pattern matching easier for microbiologists and other researchers.

8. Acknowledgment

This work was supported by grant GM048167 from the National Institutes of Health.

9. References

Bailey, T. L. & Gribskov, M. (1998). Combining evidence using p-values: application to sequence homology searches. *Bioinformatics*, Vol. 14, No. 1, pp. 48-54.

Bauer, A. L., Hlavacek, W. S., Unkefer, P. J., & Mu, F. (2010). Using sequence-specific chemical and structural properties of DNA to predict transcription factor binding sites. *PLoS Comput Biol*, Vol. 6, No. 11, pp. e1001007.

Blattner, F. R., Plunkett, G., 3rd, Bloch, C. A., Perna, N. T., Burland, V., Riley, M., Collado-Vides, J., Glasner, J. D., Rode, C. K., Mayhew, G. F., Gregor, J., Davis, N. W., Kirkpatrick, H. A., Goeden, M. A., Rose, D. J., Mau, B., & Shao, Y. (1997). The complete genome sequence of Escherichia coli K-12. *Science*, Vol. 277, No. 5331, pp. 1453-62.

Bulyk, M. L. (2003). Computational prediction of transcription-factor binding site locations. *Genome Biol*, Vol. 5, No. 1, pp. 201.

Carmack, C. S., McCue, L. A., Newberg, L. A., & Lawrence, C. E. (2007). PhyloScan: identification of transcription factor binding sites using cross-species evidence. *Algorithms Mol Biol*, Vol. 2, pp. 1.

Crooks, G. E., Hon, G., Chandonia, J. M., & Brenner, S. E. (2004). WebLogo: a sequence logo generator. *Genome Research*, Vol. 14, No. 6, pp. 1188-90.

Das, M. K. & Dai, H. K. (2007). A survey of DNA motif finding algorithms. *BMC Bioinformatics*, Vol. 8 Suppl 7, pp. S21.

D'Haeseleer, P. (2006). How does DNA sequence motif discovery work? *Nat Biotechnol*, Vol. 24, No. 8, pp. 959-61.

D'Haeseleer, P. (2006b). What are DNA sequence motifs? *Nat Biotechnol*, Vol. 24, No. 4, pp. 423-5.

Dsouza, M., Larsen, N., & Overbeek, R. (1997). Searching for patterns in genomic data. *Trends in Genetics : TIG*, Vol. 13, No. 12, pp. 497-8.

Gama-Castro, S., Salgado, H., Peralta-Gil, M., Santos-Zavaleta, A., Muniz-Rascado, L., Solano-Lira, H., Jimenez-Jacinto, V., Weiss, V., Garcia-Sotelo, J. S., Lopez-Fuentes, A., Porron-Sotelo, L., Alquicira-Hernandez, S., Medina-Rivera, A., Martinez-Flores, I., Alquicira-Hernandez, K., Martinez-Adame, R., Bonavides-Martinez, C., Miranda-Rios, J., Huerta, A. M., Mendoza-Vargas, A., Collado-Torres, L., Taboada, B., Vega-Alvarado, L., Olvera, M., Olvera, L., Grande, R., Morett, E., & Collado-Vides, J. (2011). RegulonDB version 7.0: transcriptional regulation of Escherichia coli K-12 integrated within genetic sensory response units (Gensor Units). *Nucleic Acids Research*, Vol. 39, No. Database issue, pp. D98-105.

Grote, A., Klein, J., Retter, I., Haddad, I., Behling, S., Bunk, B., Biegler, I., Yarmolinetz, S., Jahn, D., & Munch, R. (2009). PRODORIC (release 2009): a database and tool platform for the analysis of gene regulation in prokaryotes. *Nucleic Acids Research*, Vol. 37, No. Database issue, pp. D61-5.

Harlocker, S. L., Bergstrom, L., & Inouye, M. (1995). Tandem binding of six OmpR proteins to the ompF upstream regulatory sequence of Escherichia coli. *The Journal of Biological Chemistry*, Vol. 270, No. 45, pp. 26849-56.

Homann, O. R. & Johnson, A. D. (2010). MochiView: versatile software for genome browsing and DNA motif analysis. *BMC Biol*, Vol. 8, pp. 49.

Hu, J., Li, B., & Kihara, D. (2005). Limitations and potentials of current motif discovery algorithms. *Nucleic Acids Res*, Vol. 33, No. 15, pp. 4899-913.

Kondo, H., Nakagawa, A., Nishihira, J., Nishimura, Y., Mizuno, T., & Tanaka, I. (1997). Escherichia coli positive regulator OmpR has a large loop structure at the putative RNA polymerase interaction site. *Nature Structural Biology*, Vol. 4, No. 1, pp. 28-31.

Ladunga, I. (2010). An overview of the computational analyses and discovery of transcription factor binding sites. *Methods Mol Biol*, Vol. 674, pp. 1-22.

MacIsaac, K. D. & Fraenkel, E. (2006). Practical strategies for discovering regulatory DNA sequence motifs. *PLoS Comput Biol*, Vol. 2, No. 4, pp. e36.

Mahony, S. & Benos, P. V. (2007). STAMP: a web tool for exploring DNA-binding motif similarities. *Nucleic Acids Res*, Vol. 35, No. Web Server issue, pp. W253-8.

Martinez-Hackert, E. & Stock, A. M. (1997). The DNA-binding domain of OmpR: crystal structures of a winged helix transcription factor. *Structure*, Vol. 5, No. 1, pp. 109-24.

Medina-Rivera, A., Abreu-Goodger, C., Thomas-Chollier, M., Salgado, H., Collado-Vides, J., & van Helden, J. (2010). Theoretical and empirical quality assessment of transcription factor-binding motifs. *Nucleic Acids Res*, Vol. 39, No. 3, pp. 808-24.

Meysman, P., Dang, T. H., Laukens, K., De Smet, R., Wu, Y., Marchal, K., & Engelen, K. (2011). Use of structural DNA properties for the prediction of transcription-factor binding sites in Escherichia coli. *Nucleic Acids Res*, Vol. 39, No. 2, pp. e6.

Mrazek, J. (2009). Finding sequence motifs in prokaryotic genomes--a brief practical guide for a microbiologist. *Brief Bioinform*, Vol. 10, No. 5, pp. 525-36.

Mrazek, J. & Xie, S. (2006). Pattern locator: a new tool for finding local sequence patterns in genomic DNA sequences. *Bioinformatics*, Vol. 22, No. 24, pp. 3099-100.

Mrazek, J., Xie, S., Guo, X., & Srivastava, A. (2008). AIMIE: a web-based environment for detection and interpretation of significant sequence motifs in prokaryotic genomes. *Bioinformatics*, Vol. 24, No. 8, pp. 1041-8.

Munch, R., Hiller, K., Grote, A., Scheer, M., Klein, J., Schobert, M., & Jahn, D. (2005). Virtual Footprint and PRODORIC: an integrative framework for regulon prediction in prokaryotes. *Bioinformatics*, Vol. 21, No. 22, pp. 4187-9.

Novichkov, P. S., Laikova, O. N., Novichkova, E. S., Gelfand, M. S., Arkin, A. P., Dubchak, I., & Rodionov, D. A. (2010a). RegPrecise: a database of curated genomic inferences of transcriptional regulatory interactions in prokaryotes. *Nucleic Acids Research*, Vol. 38, No. Database issue, pp. D111-8.

Novichkov, P. S., Rodionov, D. A., Stavrovskaya, E. D., Novichkova, E. S., Kazakov, A. E., Gelfand, M. S., Arkin, A. P., Mironov, A. A., & Dubchak, I. (2010b). RegPredict: an integrated system for regulon inference in prokaryotes by comparative genomics approach. *Nucleic acids research*, Vol. 38, No. Web Server issue, pp. W299-307.

Oberto, J. (2010). FITBAR: a web tool for the robust prediction of prokaryotic regulons. *BMC Bioinformatics*, Vol. 11, pp. 554.

Oshchepkov, D. Y., Vityaev, E. E., Grigorovich, D. A., Ignatieva, E. V., & Khlebodarova, T. M. (2004). SITECON: a tool for detecting conservative conformational and physicochemical properties in transcription factor binding site alignments and for site recognition. *Nucleic Acids Res*, Vol. 32, No. Web Server issue, pp. W208-12.

Oshima, T., Aiba, H., Masuda, Y., Kanaya, S., Sugiura, M., Wanner, B. L., Mori, H., & Mizuno, T. (2002). Transcriptome analysis of all two-component regulatory system mutants of Escherichia coli K-12. *Molecular Microbiology*, Vol. 46, No. 1, pp. 281-91.

Palumbo, M. J. & Newberg, L. A. (2010). Phyloscan: locating transcription-regulating binding sites in mixed aligned and unaligned sequence data. *Nucleic Acids Research*, Vol. 38, No. Web Server issue, pp. W268-74.

Pavesi, G., Mauri, G., & Pesole, G. (2004). In silico representation and discovery of transcription factor binding sites. *Brief Bioinform*, Vol. 5, No. 3, pp. 217-36.

Piipari, M., Down, T. A., Saini, H., Enright, A., & Hubbard, T. J. (2010). iMotifs: an integrated sequence motif visualization and analysis environment. *Bioinformatics*, Vol. 26, No. 6, pp. 843-4.

Quest, D., Dempsey, K., Shafiullah, M., Bastola, D., & Ali, H. (2008). MTAP: the motif tool assessment platform. *BMC Bioinformatics*, Vol. 9 Suppl 9, pp. S6.

Rice, P., Longden, I., & Bleasby, A. (2000). EMBOSS: the European Molecular Biology Open Software Suite. *Trends in Genetics : TIG*, Vol. 16, No. 6, pp. 276-7.

Schug, J. (2008). Using TESS to predict transcription factor binding sites in DNA sequence. *Curr Protoc Bioinformatics*, Vol. Chapter 2, pp. Unit 2 6.

Sen, N., Mishra, M., Khan, F., Meena, A., & Sharma, A. (2009). D-MATRIX: a web tool for constructing weight matrix of conserved DNA motifs. *Bioinformation*, Vol. 3, No. 10, pp. 415-8.

Sharma, D., Mohanty, D., & Surolia, A. (2009). RegAnalyst: a web interface for the analysis of regulatory motifs, networks and pathways. *Nucleic Acids Res*, Vol. 37, No. Web Server issue, pp. W193-201.

Stormo, G. D. (2010). Motif discovery using expectation maximization and Gibbs' sampling. *Methods Mol Biol*, Vol. 674, pp. 85-95.

Tan, K., McCue, L. A., & Stormo, G. D. (2005). Making connections between novel transcription factors and their DNA motifs. *Genome Res*, Vol. 15, No. 2, pp. 312-20.

Thomas-Chollier, M., Sand, O., Turatsinze, J. V., Janky, R., Defrance, M., Vervisch, E., Brohee, S., & van Helden, J. (2008). RSAT: regulatory sequence analysis tools. *Nucleic acids research*, Vol. 36, No. Web Server issue, pp. W119-27.

Thompson, W., Conlan, S., McCue, L. A., & Lawrence, C. E. (2007). Using the Gibbs Motif Sampler for phylogenetic footprinting. *Methods Mol Biol*, Vol. 395, pp. 403-24.

Thompson, W., McCue, L. A., & Lawrence, C. E. (2005). Using the Gibbs motif sampler to find conserved domains in DNA and protein sequences. *Curr Protoc Bioinformatics*, Vol. Chapter 2, pp. Unit 2 8.

Wei, W. & Yu, X. D. (2007). Comparative analysis of regulatory motif discovery tools for transcription factor binding sites. *Genomics Proteomics Bioinformatics*, Vol. 5, No. 2, pp. 131-42.

Wittkop, T., Rahmann, S., & Baumbach, J. (2010). Efficient online transcription factor binding site adjustment by integrating transitive graph projection with MoRAine 2.0. *J Integr Bioinform*, Vol. 7, No. 3, pp. 1-11.

Yanover, C., Singh, M., & Zaslavsky, E. (2009). M are better than one: an ensemble-based motif finder and its application to regulatory element prediction. *Bioinformatics*, Vol. 25, No. 7, pp. 868-74.

Yellaboina, S., Seshadri, J., Kumar, M. S., & Ranjan, A. (2004). PredictRegulon: a web server for the prediction of the regulatory protein binding sites and operons in prokaryote genomes. *Nucleic acids research*, Vol. 32, No. Web Server issue, pp. W318-20.

Yoshida, T., Qin, L., Egger, L. A., & Inouye, M. (2006). Transcription regulation of ompF and ompC by a single transcription factor, OmpR. *The Journal of Biological Chemistry*, Vol. 281, No. 25, pp. 17114-23.

Zaslavsky, E. & Singh, M. (2006). A combinatorial optimization approach for diverse motif finding applications. *Algorithms Mol Biol*, Vol. 1, pp. 13.

Zhang, S., Xu, M., Li, S., & Su, Z. (2009). Genome-wide de novo prediction of cis-regulatory binding sites in prokaryotes. *Nucleic Acids Res*, Vol. 37, No. 10, pp. e72.

Zhou, Q. & Liu, J. S. (2004). Modeling within-motif dependence for transcription factor binding site predictions. *Bioinformatics*, Vol. 20, No. 6, pp. 909-16.

4

Understanding LiP Promoters from *Phanerochaete chrysosporium*: A Bioinformatic Analysis

Sergio Lobos[1], Rubén Polanco[2], Mario Tello[3], Dan Cullen[4], Daniela Seelenfreund[1] and Rafael Vicuña[5]

[1]*Laboratorio de Bioquímica, Departamento de Bioquímica y Biología Molecular, Facultad de Ciencias Químicas y Farmacéuticas, Universidad de Chile*

[2]*Escuela de Bioquímica, Facultad de Ciencias Biológicas, Universidad Andrés Bello*

[3] *Centro de Biotecnología Acuícola, Universidad de Santiago de Chile.*

[4]*USDA Forest Service, Forest Products Laboratory, Madison, WI 53726,*

[5] *Departamento de Genética Molecular y Microbiología, Facultad de Ciencias Biológicas, Pontificia Universidad Católica de Chile and Millennium Institute for Fundamental and Applied Biology*

[1,2,3,5]*Chile*

[4]*USA*

1. Introduction

DNA contains the coding information for the entire set of proteins produced by an organism. The specific combination of proteins synthesized varies with developmental, metabolic and environmental circumstances. This variation is generated by regulatory mechanisms that direct the production of messenger ribonucleic acid (mRNA) and subsequent translation of the nucleotide sequence into amino acid sequences, among other fundamental processes including post-translational modifications. A major step of gene expression regulation is the control of transcription initiation by RNA polymerase II. Control systems that modulate mRNA synthesis are based on the specific recognition and interaction of proteins with cognate sites on the DNA. The complex network of DNA-protein and protein-protein interactions determines the degree of transcription of a specific sequence and defines particular expression patterns. Ultimately, the outcome of this net of interactions provides the finely-tuned response to internal clues and environmental signals (Matthews, 1992).

Understanding gene expression in complex organisms such as eukaryotes is one of the most important challenges of molecular biology. One of the most fundamental and unanswered questions is whether adaptative evolution proceeds through changes in protein-coding DNA sequences or through non-coding regulatory sequences. It has been argued that morphological change occurs mainly via non-coding changes (Haygood et al., 2010). *Diptera* studies showed that *cis*-regulatory sequences that control transcription are a common source of divergent protein expression patterns and thus of phenotypic change (Wittkopp, 2006).

Also, recent analyses of the human genome suggest a distinctive role for adaptative changes both in coding and non-coding sequences. Changes in non-coding sequences appear primarily related to changes in neural development (Haygood et al., 2010).

The last decade has witnessed an explosion of studies showing that the complex regulation of gene expression is mainly modulated by the manifold interactions between transcription factors (TFs) with their corresponding transcription factors binding sites (TFBSs) on DNA (Wei & Yu, 2007). These regulatory elements are either located proximally, in sequences upstream of the transcription start site, which are generically known as promoters, or more distantly, in sequences known as enhancers or silencers. *Cis*-regulatory elements are information processing units that are embedded in genomic DNA and which regulate gene expression. Most commonly, these *cis*-regulatory elements or modules are a few dozens to several hundred base pairs long and are comprised of multiple binding sites for transcription factors. On average, a module will have binding sites for different transcription factors and for some factors, more than one site may be present (Howard & Davidson, 2004). To date, *cis*-regulatory modules of some *Drosophila* genes have been characterized at the target site level, providing an explanation of how these sequences and gene network architectures control development in early dipteran embryos (Howard & Davidson, 2004). Modules have also been denominated "motifs" by many authors, and this is the nomenclature we will use throughout this work. Knowledge of the *cis*-regulatory elements or motifs of many genes from different species may offer insight into how these sequences control the building of the diverse structures and functional adaptations found in living organisms.

As is well known, transcription involves the binding of proteins to several sites on a promoter sequence, and in eukaryotes the action of transcription factors over long distances seems to be the rule. Transcriptional outcome can be influenced by cooperative interactions of proteins between adjacent or distant sites, mainly through the formation of DNA loops, as has been described profusely in both prokaryotic and eukaryotic organisms (Han et al., 2009; Matthews, 1992; Schleif, 1992). The property of DNA to form loops enhances the regulatory properties of proteins and expands the flexibility of systems in responding to signals that evoke cellular change.

In order to understand the functional organization of a eukaryotic promoter, in this study we used the well-studied ligninolytic fungal species *Phanerochaete chrysosporium*, and examined the promoters from a selected gene family. *P. chrysosporium* has been used as a model system in numerous studies for its production of lignin-degrading enzymes (Singh & Chen, 2008). Cellulose and lignin constitute the most abundant forms of organic carbon and their degradation and mineralization is a fundamental step in the carbon cycle of the biosphere. The use of lignocellulosic biomass depends on either the removal or disruption of lignin by a process that can include the activity of lignin- and manganese-dependent peroxidases in order to expose the cellulose polymer to the attack of cellulolytic enzymes. Therefore, an understanding of the regulatory mechanisms that underlie the production of these enzymes is of pivotal importance both for a deeper comprehension of the crucial process of maintenance of the carbon cycle in nature and for the production of bioenergy. Additionally, lignocellulosic wastes are produced in large amounts and efforts have been made to convert these residues into valuable products such as biofuels, chemicals and animal feed (Dashtban et al., 2009). This bioconversion usually requires a multistep process involving a pretreatment (mechanical, chemical or biological) and hydrolysis to produce readily metabolyzable molecules such as hexoses and pentoses (Sánchez, 2009).

Pretreatment of lignocellulosic residues is necessary because hydrolysis of non-pretreated material is slow and results in low yield (Dashtban et al., 2009). It has been reported that the use of *P. chrysosporium* is advantageous for pretreatment of cotton stalks in an energy-saving, low cost and environmentally friendly approach that can reduce chemical pretreatments (Shi et al., 2009). Reported recovery depended on culture conditions, either agitated or shallow stationary submerged. Although agitated cultivation resulted in better delignification, pretreatment under submerged shallow stationary conditions provides a better balance between lignin degradation and carbohydrate availability (Shi et al., 2009). Interestingly, under solid-state cultivation, higher cellulolytic but not ligninase activity was associated with Mn^{2+} addition, although the initial purpose of supplementing Mn^{2+} was to improve ligninase activities and lignin degradation (Shi et al., 2008). This fungus has also shown promising results in wood biopulping (Singh et al., 2010) and soil bioremediation (Jiang et al., 2006). Hence, optimization of these biotechnological processes can also profit from a deeper understanding of the fundamental process of gene transcription.

Woodrotting fungi include white-rot basidiomycetes, brown-rot basidiomycetes, and soft-rot ascomycetes/deuteromycetes; however, only a small group of these are able to completely degrade lignin to carbon dioxide and thereby gain access to the carbohydrate polymers of plant cell walls, which they use as carbon and energy sources. Selective degradation of lignin by these fungi leaves behind crystalline cellulose with a bleached appearance that is often referred to as "white rot" (Martínez et al., 2004). Some or all of these enzymes and their isozymes of the lignin depolymerization system include multiple isozymes of lignin peroxidase (LiP) and manganese-dependent peroxidase (MnP) (Kirk & Farrell, 1987; Farrell et al., 1989; Singh & Chen, 2008). Among the ligninolytic fungi, *P. chrysosporium* is considered as a model organism for the development and understanding of the ligninolytic-enzyme-production system, as it can produce a more complete ligninolytic enzyme complex than most other species (Kirk & Farrell, 1987) and until recently, it was the only ligninolytic fungus whose genome has been sequenced (Martínez et al., 2004). In *P. chrysosporium*, LiPs together with MnPs and H_2O_2-producing enzymes constitute the major components of the lignin-degrading system that are secreted to the extracellular medium (Kirk & Farrell, 1987; Farrell et al., 1989; Kirk et al., 1990).

The characterization of ligninolytic enzyme systems of several basidiomycetes has revealed that in some species LiP activity is not observed. For example, in the white rot fungus *Phanerochaete sordida* only MnP activity, but no lignin peroxidase or laccase activity was detected, although several culture conditions were assayed. In this species, three highly similar MnP isoenzymes were identified (Rüttimann-Johnson et al., 1994). The white-rot basidiomycete *Ceriporiopsis subvermispora* produces two families of ligninolytic enzymes, MnPs and laccases (Lobos et al., 1994), but lignin peroxidase activity is not detected (Rajakumar et al., 1996). In *Ganoderma lucidum* low levels of MnP activity are detected in some culture media, but not in others and no LiP activity was seen in any of the media tested (D'Souza et al., 1999).

The genome of *P. chrysosporium* contains a large group of genes coding for low-redox peroxidases (LRP), including 10 *lip* genes, 5 genes coding for MnPs, 4 genes encoding multicopper copper oxidases (related to laccases) and an interesting peroxidase gene unlinked to all peroxidases, that shares residues common to both MnPs and LiPs (Martínez et al., 2004). Other white rot fungi, such as *C. subvermispora* (Rajakumar et al., 1996) and G. *lucidum* (D'Souza et al., 1999) also contain *lip*-like genes, but as described above, do not exhibit detectable LiP activity.

The recent genome sequencing of a second basidiomycete, the brown-rot fungus *Postia placenta,* yielded exciting novelties: genes encoding the class II secretory peroxidases LiP, MnP and versatile peroxidase were not detected in the *P. placenta* genome (Martínez et al., 2009). This fungus contains only one LRP gene that is not closely related to LiP and MnP, but is part of an assemblage of "basal peroxidases" that includes the novel peroxidase (NoP) of *P. chrysosporium* (Martínez et al., 2009). Comparison of the *P. placenta* and *P. chrysosporium* genomes indicates that the derivation of brown-rot is characterized largely by the contraction or loss of multiple gene families that are thought to be important in typical white-rot, such as cellulases, LiPs, MnPs, copper radical oxidases, among other enzymes. Phylogenetic analysis suggests that LiP and MnP gene lineages of *P. chrysosporium* were independently derived from the basal peroxidases before the divergence of *Postia* and *Phanerochaete.* If so, then the absence of LiP and MnP in *P. placenta* may reflect instances of gene loss (Martínez et al., 2009). This general pattern of simplification is consistent with the view that brown-rot fungi, having evolved novel mechanisms for initiating cellulose depolymerization, have cast off much of the energetically costly lignocellulose-degrading apparatus that is retained in white-rot fungi, such as *P. chrysosporium* (Martínez et al., 2009).

LiPs from *P. chrysosporium* are encoded by ten structurally related genes (Stewart & Cullen, 1999). The genomic organization of the *lip* genes that encode these isoenzymes is known: four genes (*lipA, lipB, lipC* and *lipE*) reside within a 35 Kb region and the remaining genes (*lipG, lipH, lipI* and *lipJ*) lie within a 15 Kb region, forming clusters where six genes occur in pairs that are transcriptionally convergent (Stewart & Cullen, 1999). The transcriptional orientation and intergenic distances indicate that regulatory promoter sequences are not shared among any of the *lip* genes. *Lip* genes have been classified by their deduced amino acid sequences and also by their intron/exon structure (Stewart & Cullen, 1999). The phylogenetic clustering defines a major subfamily I of six genes (*lipA, lipB, lipE, lipG, lipH* and *lipI*) and four minor subfamilies of only one member each (*lipC, lipD, lipF* and *lipJ*) (Stewart & Cullen, 1999).

Although the *lip* genes are structurally related and the proteins participate in a common physiological process, *lip* promoter sequences display no obvious similarities, suggesting differential gene expression of this family of isozymes. Indeed, the relative transcriptional activity of these genes has been assessed systematically, showing differential regulation in response to carbon (C)-limited or nitrogen (N)-limited culture media (Stewart & Cullen, 1999). Recently, it was shown that over a hundred proteins that are secreted by *P. chrysosporium* exhibited increased transcription in either C- or N-limited relative to nutrient replete medium, including LiP and MnP expression (Wymelenberg et al., 2009). In another study, similar expression patterns of secreted proteins between cellulose-grown and wood-grown cultures were found (Sato et al., 2007), but this study showed the complication of considering wood as a nutrient, since it is both N-limited and C-replete. In addition to enzymes which act on lignocelluloses, proteases were found, suggesting the ability to generate nitrogen (Sato et al., 2007); depletion of nitrogen triggers the onset of secondary metabolism. Metabolic switching occurs in culture after 48 hours when linear growth ceases. After 72 hours, *P. chrysosporium* has shifted to secondary metabolism, its beginning being closely related to the appearance of LiP activity (Wu & Zhang, 2010). The complex expression pattern of *lip* genes suggests that each isozyme might play a specific biological role in the process of ligninolysis, though why there is a multiplicity of lignin peroxidases remains unclear (Farrell et al., 1989; Stewart & Cullen, 1999; Sato et al., 2007). This long standing question is especially intriguing and paradoxical, since LiPs are low-redox enzymes that catalyze a unique nonspecific enzymatic "combustion", i.e. susceptible aromatic substrate molecules are oxidized by one electron and this produces unstable cation radicals which then undergo a variety of nonenzymatic reactions (Kirk & Farrell, 1987). The answer to this fundamental issue is still a matter of debate and it is speculated that an array of different genes may provide the necessary plasticity to the fungus to attack diverse types of lignin, its recalcitrant carbon and energy source, under various biotic and abiotic conditions.

The isoenzyme family of LiP proteins from *P. chrysosporium* provides an interesting model for analyzing the evolution of promoters and their coding sequences. The identification of characteristic features regulating the main genes involved in lignin biodegradation, as well as others that are co-regulated, can both provide a more complete understanding of promoter organization and be used to identify novel genes involved in ligninolysis through bioinformatics-based searches. In this study, both bioinformatic tools and experimental data were used to explore if the structure of promoter organization is related to the phylogenetic grouping of the LiP proteins. A motif is a pattern common to a set of nucleic acid subsequences which *share some biological property of interest,* such as being a DNA binding site for a regulatory protein. It was expected that these motifs would provide information about the regulatory factors that control gene expression and identify transcription factors that bind to the motifs. The main goal was to analyze the structural organization of the promoters of the *lip* gene family and determine if there exists an organization of TFBSs and/or some kind of structured assembly of *cis*-regulatory elements or motifs within their promoter sequences. The promoter structures were compared with reported data on the differential regulation, transcription and phylogenetic analysis of the LiP proteins. To our knowledge, no reports exist where bioinformatic data has been correlated with the expression of a family of isoenzymes in filamentous fungi. The working hypothesis of this study was to establish if genes involved in the same biological process have promoters that share structural characteristics, although these common structural elements may not be evident. In this case, it should be possible detect a common architecture using appropriate bioinformatics tools in order to identify motif patterns that contain functional TFBSs.

2. Method and results

2.1 Analysis of promoters from lignin peroxidase genes

2.1.1 Alignment of promoter sequences reveals a similar pattern to *lip* gene clustering

We first analyzed 1 Kb of the available promoter sequences of the ten *lip* genes using the ClustalW (Thompson et al., 1994) and the Jotun Hein algorithms (Hein, 1990) from the DNAStar software (Figure 1). Due to lack of information about the transcriptional start sites of the *lip* genes from *P. chrysosporium,* we first analyzed 1 Kb of the available region located upstream of the translational start site, since it was highly probable that promoter sequences were included. We tested two alignment algorithms: one was the Needleman-Wunsch algorithm present in ClustalW, which does not presume an evolutionary relationship between the analyzed sequences. The second was the Jotun Hein algorithm, a Markov chain algorithm that presumes an evolutionary relationship between the sequences to be analyzed. ClustalW was performed using BLOSUM62 matrix. The algorithms were used through the PC interphase provided by the DNASTART software. When using the Jotun Hein algorithm, a clustering of the promoters belonging to the subfamily I *lip* genes appeared (i.e. *lipA, lipB, lipE, lipG, lipH* and *lipI*), that is similar to the relationship between the protein sequences (Stewart & Cullen, 1999). Cladistic analysis based on promoter sequences showed two main branches within the family. The main branch included all but one of the promoter sequences of the *lip* family I, conformed by *lipA, lipB, lipE, lipG, lipH* and *lipI*. The sole exception was *lipH*, which appeared more closely linked to the *lipF* promoter sequence. The sequences corresponding to the promoters of *lipD, lipC* and *lipJ*, which comprise the subfamilies II, III and V, respectively, were more divergent. This grouping was also apparent when the ClustalW analysis was repeated using 2 Kb of all promoter sequences. Both algorithms were able to detect an evolutionary relationship between upstream regions of the *lip* genes, but the Jotun Hein algorithm was more sensitive to detect this relation. The fact that very similar results were obtained using two different algorithms, suggests that this association is not spurious and supports the finding of a common organization of the analyzed sequences. Jotun-Hein was also used because it had been employed for analysis of the LiP proteins (Stewart & Cullen, 1999).

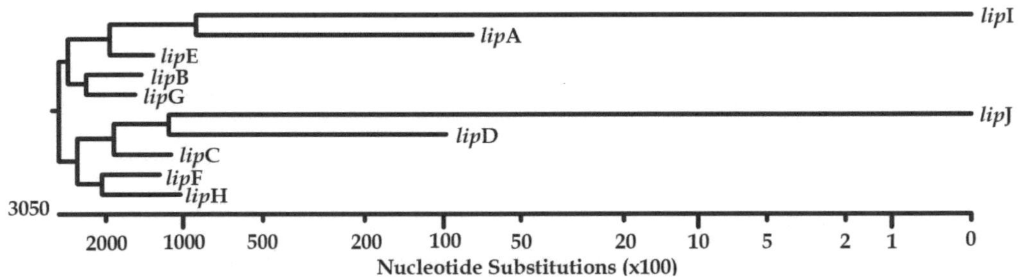

Fig. 1. Cladistic analysis of 1 Kb promoter sequences of 10 *lip* (lignin peroxidase) genes from *Phanerochaete chrysosporium*. Each sequence in the analysis corresponds to 1 Kb upstream of the ATG codon. Analysis was performed with the Jotun-Hein (Hein, 1990) algorithm on LASERGENE package software.

When the Clustal analysis included only the six promoter sequences of the genes belonging to the subfamily I of *lip* genes, a similar order appeared where *lipH* again corresponded to the most distant member of the group (Figure 2).

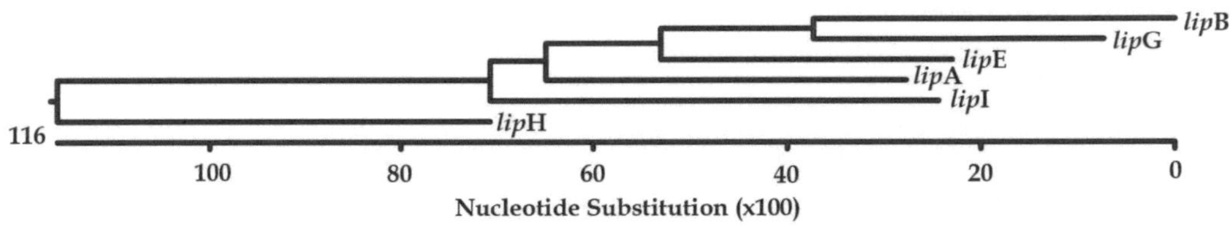

Fig. 2. Cladistic analysis of 1 Kb of six *lip* promoters corresponding to the Subfamily I classification from *Phanerochaete chrysosporium*. Each sequence in the analysis corresponds to 1 Kb upstream of the ATG codon. Analysis was done with the ClustalW (Thompson et al., 1994) algorithm on DNAStar software.

2.1.2 Defining an ATG upstream region for analysis

We then analyzed the available ATG upstream region of the ten *lip* genes using the Genomatix bioinformatics tool that searches conserved *cis*-regulatory elements within

TRANSFAC and JASPAR databases. It is not possible to precisely define promoter sequences, as the transcription start site is unknown in this case and it is not easy in general to define how far upstream distal sequences control gene expression. Therefore, sequences upstream of the ATG of 500, 1000 and 2000 bp were analyzed for the presence of conserved *cis*-regulatory elements or TFBSs. With this tool, a multiplicity of elements was evident; however, no clear pattern of structural organization emerged. Thus, a more sophisticated method to find sequence patterns was needed. Among programs that perform this kind of analysis, MEME (Multiple Expectation maximization for Motif Elicitation) and Gibbs are two well-documented programs for this purpose. We chose MEME because the algorithm for maximation of Multiple Expectation allows defining more clearly a motif pattern independent of its position in the sequence. On the other hand, TRANSFAC and JASPAR allow the identification of putative binding sites only for known transcription factors, but do not find new regulatory elements, especially in organisms that have not been extensively studied. When upstream sequences (500, 1000 and 2000 bp) were analyzed using MEME software, a pattern of elements emerged that split the *lip* promoters into two groups, where the genes of one group again corresponded to the members of subfamily I of *lip* genes. This separation was subtle when analyzing 500 bp or 2000 bp of the promoter sequence but was more evident when analyzing 1000 bp of the regulatory sequences. An additional reason for choosing 1 Kb ATG upstream sequences for analysis is that, as explained above, transcriptional outcome can be influenced by cooperative interactions of proteins, mainly through the formation of DNA loops. This looping depends on the probability of two sites coming together, which is optimal for cyclization at 500 bp and decreases at distances greater than 1000 bp (Matthews, 1992). For these reasons, a promoter size of 1000 bp was chosen for further studies.

2.2 Analytical strategy to identify regulatory elements

The next step consisted of applying a set of analytical tools to identify putative regulatory elements within the *lip* gene family. In a step-wise strategy, first putative motifs were identified with MEME; then, for each motif, integrated databases were searched for genes that contained these motifs in their promoters with the MAST software. Briefly, MAST takes any motif and transforms it into a position-dependent scoring matrix that is used to scan a curated database of promoter sequences. Finally, to identify if this sequence corresponds to a transcriptional binding site, the best match obtained in the yeast database is used by MAST to screen a transcription factors database in order to identify the TF that recognizes the yeast sequence with the motif identified by MEME. The database used for this purpose was YEASTEXTRACT. To summarize, a general streamlined approach was defined to identify a putatively functional structure in eukaryotic promoters, as outlined in Figure 3. The flowchart shows the pathway for the identification of putative motifs, TFBSs and transcription factors involved in the expression of genes containing such motifs. With this analysis, five conserved motifs were identified and characterized in the promoters of *lip* genes from *P. chrysosporium*.

2.2.1 Discovery of motifs within promoters

The search for signals within the DNA sequence was carried out using MEME (Multiple EM for Motif Elicitation), a tool that was designed to discover signals (called motifs) within a set of sequences believed to share some common (but unknown) property, such as binding sites

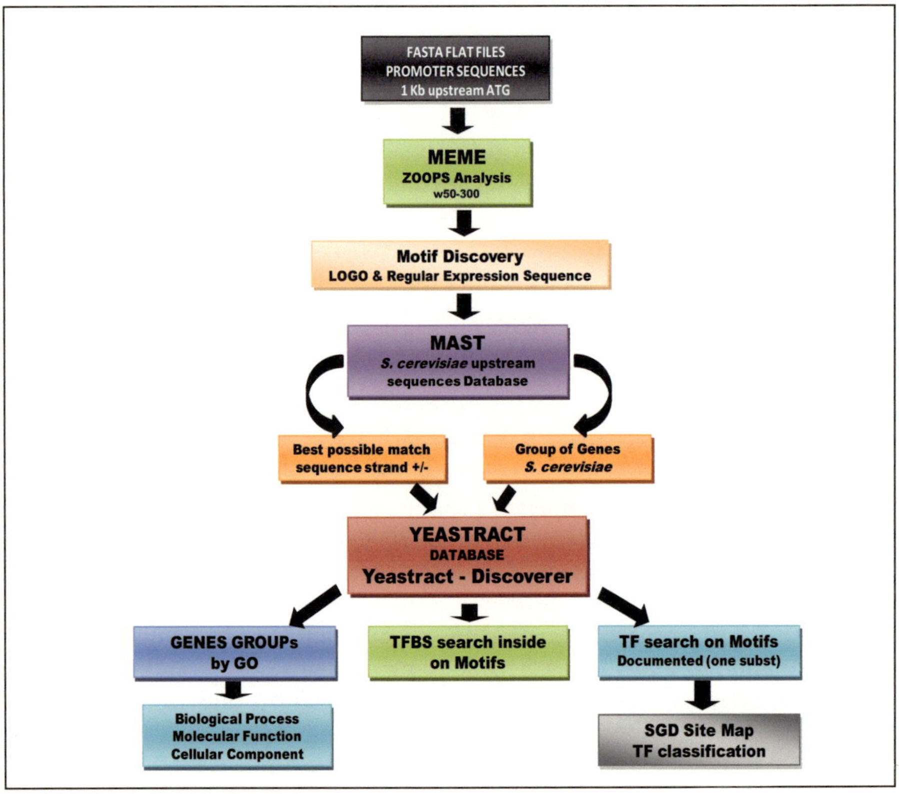

Fig. 3. Flowchart for identifying and testing putative motifs and transcription factors involved in gene expression of *lip* genes. Algorithms used at each stage are discussed in the text.

for shared transcription factors or TFBSs in a set of promoters (Bailey et al., 2006). Expectation-maximization (EM) algorithm is a method for finding maximum likelihood or maximum *a posteriori* estimates of parameters in statistical models, where the model depends on unobserved latent variables. EM is an iterative method which alternates between performing an expectation (E) step, which computes the expectation of the log-likelihood evaluated using the current estimate for the latent variables, and a maximization (M) step, which computes parameters maximizing the expected log-likelihood found on the E step. These parameter-estimates are then used to determine the distribution of the latent variables in the next E step (Dempster et al., 1977). By default, MEME assumes that every position in every sequence is equally likely *a priori* to be a motif site and can search for DNA motifs on either strand (Bailey et al., 2010). MEME finds motifs by identifying highly correlated stretches of letters in the input sequences and applies statistical models to validate the most significant motifs contained in these input sequences. Finally, it reports an E-value for each motif, giving a measure of the motif's validity or likelihood of not being a random sequence artifact (Bailey & Elkan, 1994). MEME can be accessed at the web server hosted at the http://meme.ncbr.net site and is preferentially set for searching motifs within sequences of 1 Kb (Bailey et al., 2006).

A TFBS is defined as a conserved, relatively short sequence element of 10-15 bp (Stepanova et al., 2005). Since TFBSs tend to be short and degenerate, the discovery of these sequences is a difficult task. The motif discovery algorithm searches for a minimum of two elements of similar short sequences of at least 6 bp; these motifs are searched within sliding window frames of 6 to 300 bp of width (Bailey et al., 2006). We therefore searched for motifs performing a serial analysis using 15-300, 20-300, 50-300, 100-300, 150-300, 200-300, 250-300 and 300-300 bp frames. The analysis was performed for all 10 *lip* promoters and showed a conserved pattern of motifs (Figure 4). This analysis produced a readily apparent structural organization of the *lip* genes. The motifs were most clearly noticeable with frames 15-300, 20-300 and 50-300 bp and declined with wider frames. For this reason, all further analyses were performed using the 50-300 bp window frame.

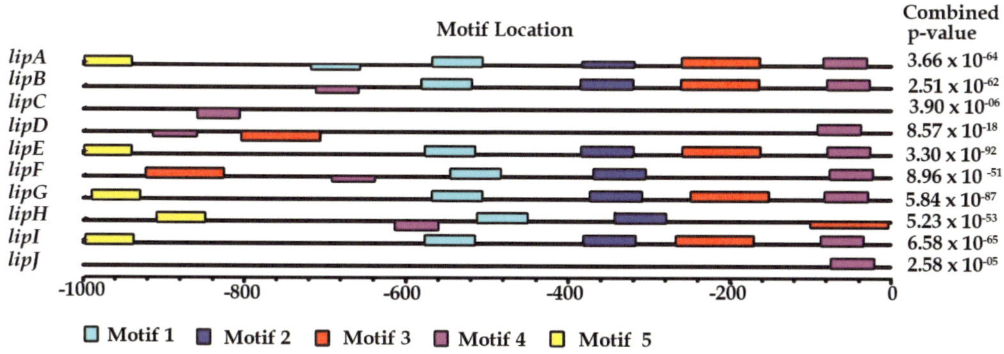

Fig. 4. The 5 most conserved motifs of the *lip* genes promoters. Maximum number of motifs: 5; windows for each motif from 50 to 300 bp. All other parameters of the MEME software corresponded to the default setting.

MEME analysis was performed using FASTA flat-files. Files containing 1 Kb of the promoter sequence from each of the 10 genes were aligned and searched for motifs with ZOOPS (Zero Or One Per Sequence) analysis. A pattern of five motifs emerged, which also corresponds to the maximum number of motifs allowed when using a 50-300 bp window frame (Figure 4). As a control, the same analysis was conducted with the promoter sequences from the subfamily I genes; when only the six promoters of the subfamily I genes were aligned, a most striking pattern of motifs emerged. Using 6-300, 20-300, 50-300 and 100-300 bp frames maintained the conspicuous pattern of five motifs that clearly indicated again a conserved organization of all six members of the subfamily I *lip* genes (Figure 5).

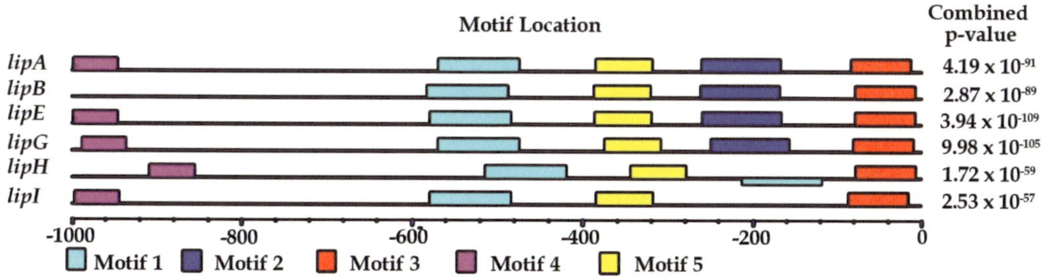

Fig. 5. Summary of the 5 most conserved motifs of the subfamily I of *lip* genes promoters. Maximum number of motifs: 5; windows per each motif from 50 to 300 bp. All other parameters of the MEME software were the default setting.

Analysis was also done using the 50-300 bp window frame or motif width window (number of characters in the sequence pattern), since the five motifs obtained presented significant E-values ranging from e-018 to e-003, which also corresponded to the best E-values of all analyzed sliding windows. Cut off E-values were set at e-003. The obtained motifs corresponded to ambiguous regular expression (LOGO) sequences of 50 to 92 bp for the five motifs identified in the subfamily I *lip* promoters and between 50 and 94 bp for the promoters of the ten *lip* genes using the MEME algorithm. To determine if the motifs found were statistically significant, the sequences were shuffled and compared to the former (training) set. Analysis of the shuffled sequences revealed that the observed motifs and the statistical significance were lost. Therefore, the structuring found in the promoter sequences was not trivial and possibly corresponds to a functional organization.

2.2.2 Analysis of motifs using MAST

In order to illustrate the effectiveness of the proposed strategy, as outlined in the flow sheet shown in Figure 3, analysis of the promoters of the subfamily I *lip* genes is described. The next step consisted in analyzing MEME results (LOGO) and regular expression sequences for the five motifs using MAST (Motif Alignment and Search Tool), which searches promoter motifs (best possible matches) in wide upstream sequences available in different databases. As mentioned before, MAST uses a position-dependent scoring matrix to search in a sequence for a segment with the best match. To perform this, MAST transforms any sequence pattern (motif) into a position-dependent scoring matrix. This means that a position-dependent scoring matrix is not applied to the end of a sequence or if any gap is present. The sequences are ranked according to their E-values. MAST searches databases for sequences that match the motifs and outputs detailed annotation showing genes that contain these motifs (Bailey & Gribskov, 1998) (Figure 6).

The findings of MAST in a particular upstream sequence database allowed obtaining a group of genes containing particular motifs in their promoters. The most comprehensive eukaryotic promoter databases are human, *Drosophila* and mouse; however, considering the relative phylogenetic closeness to the model fungal species *P. chrysosporium*, a yeast database was used for the analysis. The *Saccharomyces cerevisiae* genome database (SGD) is a repository of organized collection of yeast proteins and genes and their corresponding regulatory sequences and is probably the most appropriate database available today for fungal species. Using the SGD, the best possible match was found for motifs on either strand of each promoter. The obtained matches corresponded to defined and unambiguous sequences for the five motifs identified using the MEME algorithm. Sequences were subjected to MAST analysis for each separate motif and were also analyzed when combined. The best combined matches were found for these five motifs with varying E-values: motifs 1, 2 and 3 exhibited E-values of e-005, motifs 4 and 5 presented values of e-003 and e-002, respectively. The number of genes that contained the identified motifs that contained the identified motifs varied from five for motif 1, to 14 or 16 genes for motifs 2, 3 and 4 and enlarged to 23 genes for motif 5 (Table 1).

2.2.3 Analysis of conserved TFBSs inside each motif

Once the yeast genes that share the motif found in ATG upstream sequences of *lip* genes were obtained, transcriptional factors that bind to these sequences were analyzed with

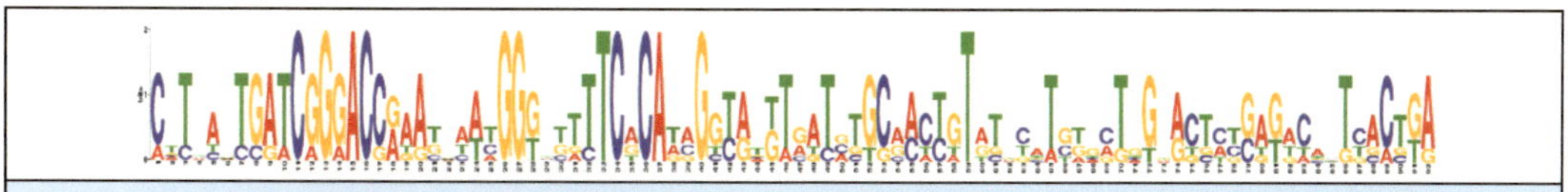

Motif 1. E-value: 1.8e-019; Width:92

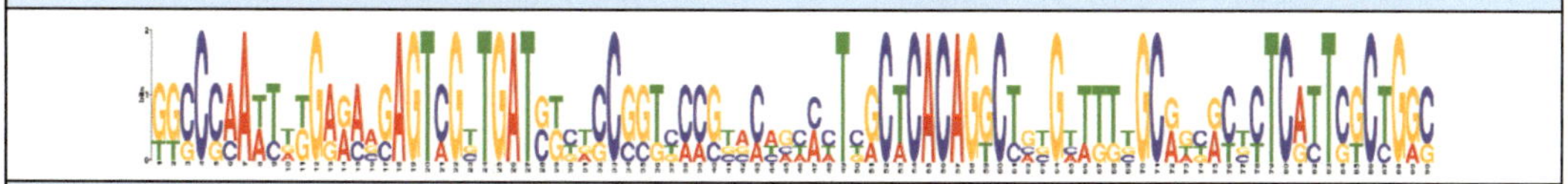

Motif 2. E-value: 1.2e-008; Width: 90

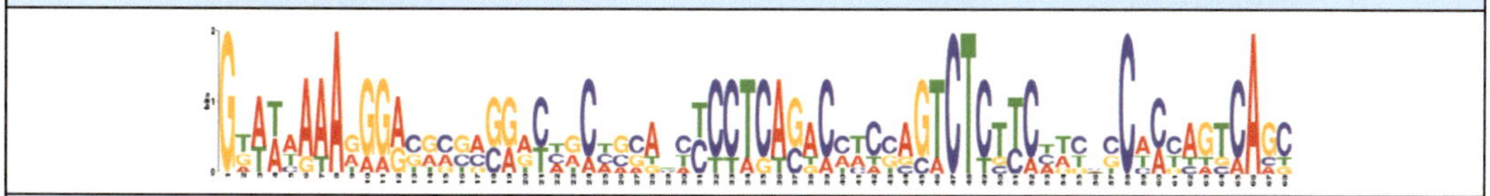

Motif 3. E-value: 3.4e-006; Width: 68

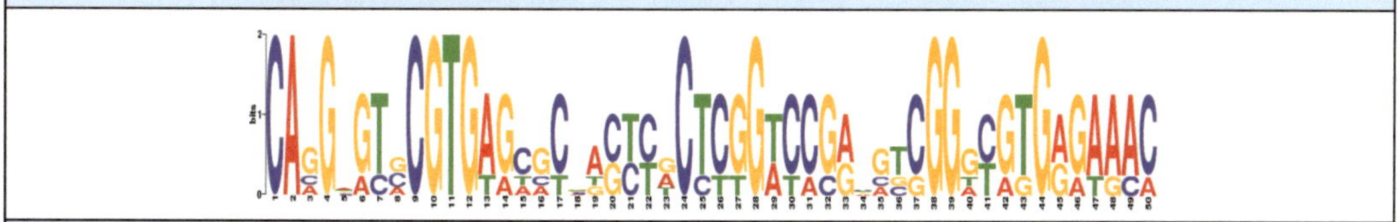

Motif 4. E-value: 1.2e-005; Width: 50

Motif 5. E-value: 1.4e-003; Width: 63

Fig. 6. LOGO representation of Motifs 1 to 5 from promoter sequences of the subfamily I *lip* genes.

YEASTRACT-DISCOVERER (YEAst Search for Transcriptional Regulators And Consensus Tracking; http://www.yeastract.com), a tool developed to support the analysis of transcription regulatory associations in yeast which can be used to identify complex motifs over-represented in promoter regions of co-regulated genes (Monteiro et al., 2008). This database contains over 48.000 documented regulatory associations between transcription factors (TFs) and target genes (Abdulrehman et al., 2011), and includes 284 specific DNA binding sites for 108 characterized TFs (Monteiro et al., 2008). To identify TFBS inside of the motifs, Yeastract uses the Smith–Waterman algorithm that allows local alignments between sequences (Smith & Waterman, 1981). The pattern matching method of YEASTRACT in search of TFBSs leads to the identification of putative target genes for specific TFs (Monteiro et al., 2008). The SGD database was therefore used to find yeast genes containing the motifs identified within the promoter sequences of the six genes of the subfamily I *lip* genes from *P. chrysosporium*.

For each motif (1 to 5), a list of yeast genes was identified. The genes that contained one of these motifs in their promoter sequence and that are included in the YEASTRACT database were further analyzed. Each gene was queried using the SGD and finally searched with GO (Gene Onthology) and its nature determined according to three defined categories: Biological process, molecular function and cellular component (Table 2).

<table>
<tr><th colspan="5">Motifs 1-5 Subfamily I Group of Genes</th></tr>
<tr><th>Motif 1 (+) | (-)
Evalue: 1e-05</th><th>Motif 2 (+) | (-)
Evalue: 1e-05</th><th>Motif 3 (+) | (-)
Evalue: 1e-05</th><th>Motif 4 (+) | (-)
Evalue: 1e-03</th><th>Motif 5 (+) | (-)
Evalue: 1e-02</th></tr>
<tr>
<td>YGR209C | TRX2
YLR173W | YLR173W
YPR094W | RDS3
YLR246W | ERF2
YIR042C | YIR042C</td>
<td>YHR135C | YCK1
YPL069C | BTS1
YDR101C | ARX1
YKL085W | MDH1
YGL047W | ALG13
YER039C-A | YER039C-A
YER040W | GLN3
YGL186C | TPN1
YIL071C | PCI8
YER038C | KRE29
YJR148W | BAT2
YBL066C | SEF1
YKL065C | YET1
YKL064W | MNR2</td>
<td>YOL132W | GAS4
YOR034C-A | YOR034C-A
YDR477W | SNF1
YPR181C | SEC23
YPR182W | SMX3
YDR059C | UBC5
YDR060W | MAK21
YHR112C | YHR112C
YHR113W | YHR113W
YOL162W | YOL162W
YDR316W | OMS1
YGR094W | VAS1
YNL268W | LYP1
YOR042W | CUE5
YMR303C | ADH2
YMR304W | UBP15</td>
<td>YLR194C | YLR194C
YJL006C | CTK2
YDL174C | DLD1
YDL173W | YDL173W
YKL020C | SPT23
YKL019W | RAM2
TS(CGA)C | SUP61
YML123C | PHO84
YML121W | GTR1
YML028W | TSA1
YBR170C | NPL4
YNL033W | YNL033W
YBR268W | MRPL37
YLR297W | YLR297W</td>
<td>YFR021W | ATG18
YGR098C | ESP1
YCR084C | TUP1
YCR086W | CSM1
YKR024C | DBP7
YKR025W | RPC37
YLR327C | TMA10
TS(GCU)L | TS(GCU)L
TG(GCC)P2 | TG(GCC)P2
TY(GUA)F2 | SUP6
YFR028C | CDC14
YPL103C | FMP30
YPL101W | ELP4
YOR180C | DCI1
YOR181W | LAS17
YDR178W | SDH4
YOR140W | SFL1
YDL079C | MRK1
YCR076C | YCR076C
YKL041W | VPS24
TY(GUA)M1 | SUP5
YBR068C | BAP2
YGR274C | TAF1</td>
</tr>
<tr><td>5</td><td>14</td><td>16</td><td>14</td><td>23</td></tr>
</table>

Table 1. Group of genes found in *S. cerevisiae* that share TFBSs found in Motifs 1 to 5.

The information of "Cellular component" for each gene was retrieved directly from the SGD database for every individual gene identified in the previous step. YEASTRACT simultaneously searches for TFBSs contained in each motif found and also searches for documented TFs that bind to these motifs (See Figure 3). This approach reduces output to a tractable size, amenable to different kinds of analysis (Table 2).

Putative functions of the identified genes suggest an interesting grouping: Motif 1 includes a single gene (Trx2) involved in cellular response to oxidative stress that presents electron carrier activity. It is noteworthy that the gene Trx2 corresponds to a cytoplasmic thioredoxin isoenzyme that is present in fungal cell walls. Motif 2 is found mainly in genes related to nitrogen metabolism and protein biosynthesis and appears to participate in biological processes of cell aging. Several of these genes are involved in biosynthetic processes of amino acids, amines and isoprenoids and also in the catabolism of amino acids. Motif 3 seems to be related to biological processes of cellular response to nitrogen and carbon metabolism and possibly, growth and differentiation. Genes containing this motif are involved in catabolic processes and cell aging, including cellular response to nitrogen starvation and eventually fungal cell wall assembly. Motif 3 is the most proximal motif identified and includes the TATA-box. This *cis*-element is conserved in all members of the subfamily I lip genes and also in all members of the *lip* gene family (in Figure 4 it corresponds to motif 4, the most proximal regulatory element for all genes, with the exception of *lip*C). Indeed, the TATA-box is conserved in approximately 30% of all eukaryotic genes (Mariño-Ramírez et al., 2004) and therefore might correspond to an ancestral regulatory feature. TATA element recognition has remained constant over the course of evolution. Genes encoding TATA-binding proteins (TBPs) have been cloned from organisms ranging from archaea to human and all share a phylogenetically conserved 180-residue carboxyterminal or core segment, which supports all of the protein's biochemically important functions in RNA Polymerase II transcription (Patikoglou et al., 1999). Motif 4 is present in several genes that do not seem to relate to a common biological process. However, one of these is an ion transporter. The finding of this motif in a gene coding for a manganese/phosphate transporter is specially striking, since MnPs also participate in lignin catabolism. This motif corresponds to the most distal element in the studied promoter (see Figure 5). Motif 5 is related to mitosis, cell cycle, chromosome segregation and stress response. The relevance of genes associated with each motif will be discussed below.

Since all genes analyzed were identified in the yeast database (SGD), an important consideration was to determine if orthologous genes exist in the genome of *P. chrysosporium*. A preliminary search in the genome of this basidiomicete (Martínez et al., 2004) indicated that all genes shown in Table 2, with the exceptions of ARX1, MDH1, SPT23, GTR1 and DLD1, are present in the *P. chrysosporium* genome.

MEME	MAST AND YEASTRACT ANALYSIS	*Saccharomyces cerevisiae* Genome Database GO ANNOTATIONS BIOLOGICAL PROCESS	MOLECULAR FUNCTION	CELLULAR COMPONENT
MOTIF 1	YGR209C | TRX2	Cell redox homeostasis - cellular response to oxidative stress ER to Golgi vesicle-mediated transport Retrograde vesicle-mediated transport, Golgi to ER Vacuole fusion (non-autophagic) - vacuole inheritance Protein deglutathionylation Regulation of DNA replication Sulfate assimilation	Disulfide oxidoreductase	Cytosol Fungal-type vacuole
MOTIF 2	YJR148W | BAT2	Branched chain family amino acid biosynthetic process Branched chain family amino acid catabolic process	Branched-chain-amino-acid transaminase	No manually curated
	YPL069C | BTS1	Terpenoid biosynthetic process	Farnesyltranstra-nsferase	No manually curated
	YKL085W | MDH1	Aerobic respiration - Tricarboxylic acid cycle Chronological cell aging - Replicative cell aging	L-malate dehydrogenase	Mitochondrion - mitochondrial matrix
	YDR101C | ARX1	Ribosomal large subunit biogénesis	Unknown	Cytoplasm - Colocalizes with cytosolic large ribosomal subunit - Nucleoplasm
	YBL066C | SEF1	Unknown	Unknown	Unknown
	YER040W | GLN3	Nitrogen catabolite activation of transcription	Sequence-specific DNA binding transcription factor	Cytosol - Nucleus

MOTIF 3	YMR303C | ADH2	Amino acid catabolic process to alcohol via Ehrlich pathway Ethanol metabolic process – Fermentation NADH oxidation	Alcohol dehydrogenase (NAD) activity	No manually curated
	YDR477W | SNF1	Cellular response to nitrogen starvation Invasive growth in response to glucose limitation Pseudohyphal growth Regulation of carbohydrate metabolic process Positive regulation of gluconeogenesis Biofilm formation - Cell adhesion Protein phosphorylation - Signal transduction Negative regulation of translation Replicative cell aging	AMP-activated protein kinase activity	AMP-activated protein kinase complex Cytoplasm Fungal-type vacuole Nuclear envelope lumen Nucleus
	YPR182W | SMX3	Nuclear mRNA splicing, via spliceosome	Unknown	U1 snRNP U4/U6 x U5 tri-snRNP complex U5 snRNP
	YGR094W | VAS1	Valyl-tRNA aminoacylation	Valine-tRNA ligase activity	Cytoplasm Mitochondrion
	YOL132W | GAS4	Ascospore wall assembly	1,3-Beta-glucanosyltransferase	Fungal-type cell wall
MOTIF 4	YKL020C | SPT23	Fatty acid metabolic process Positive regulation of transcription from RNA polymerase II promoter Response to cold	Transcription activator activity	Integral to endoplasmic reticulum membrane Nucleus
	YML121W | GTR1	Chromatin silencing at telomere Phosphate transport Transcription from RNA polymerase I / III promoters	GDP / GTP binding	Cytoplasm -GSE complex Late endosome membrane Nucleus - Vacuolar membrane
	YBR268W | MRPL37	Mitochondrial translation	Structural constituent of ribosome	Mitochondrial large ribosomal subunit
	YDL174C | DLD1	Aerobic respiration Cellular carbohydrate metabolic process	D-lactate dehydrogenase (cytochrome) activity	Mitochondrial inner membrane Mitochondrion
	YML123C | PHO84	Manganese ion transport Phosphate transport - Polyphosphate metabolic process	Inorganic phosphate transmembrane transporter activity Manganese ion transmembrane transporter activity	Integral to plasma membrane
MOTIF 5	YFR028C | CDC14	Mitotic cell cycle Nucleolus organization Protein dephosphorylation Regulation of exit from mitosis	Phosphoprotein phosphatase	Nucleolus RENT complex Spindle pole body
	YGR098C | ESP1	Apoptosis Mitotic sister chromatid segregation Negative regulation of protein phosphatase type 2A activity Regulation of exit from mitosis Regulation of mitotic spindle elongation	Cysteine-type endopeptidase	Cytoplasm Nucleus Spindle
	YDR178W | SDH4	Cellular respiration Mitochondrial electron transport,	Contributes to succinate	Mitochondrial respiratory chain

MOTIF 5		succinate to ubiquinone Tricarboxylic acid cycle	dehydrogenase (ubiquinone) activity	complex II
	YOR180C | DCI1	Not clearly defined	Not clearly defined	Peroxisomal matrix
	YOR181W | LAS17	Actin cortical patch localization Actin filament organization / actin polymerization or depolymerization Positive regulation of actin filament bundle assembly Bipolar cellular bud site selection Cytokinesis - Endocytosis Response to osmotic stress	Cytoskeletal protein binding	Actin cortical patch
	YOR140W | SFL1	Negative regulation of transcription from RNA polymerase II promoter	Specific transcriptional repressor activity Specific RNA polymerase II transcription factor activity Transcription activator activity	Nuclear chromosome
	YKR024C | DBP7	Ribosomal large subunit assembly rRNA processing	ATP-dependent RNA helicase activity	Nucleolus
	YDL079C | MRK1	Protein phosphorylation Regulation of protein catabolic process Response to stress	Protein serine/threonine kinase activity	Unknown
	YGR274C | TAF1	Gene-specific transcription from RNA polymerase II promoter General transcription from RNA polymerase II promoter RNA polymerase II transcriptional preinitiation complex assembly	Chromatin binding RNA polymerase II transcription factor activity Histone acetyltransferase activity Protein complex scaffold TATA-binding protein binding	Transcription factor TFIID complex
	YKL041W | VPS24	Intralumenal vesicle formation Late endosome to vacuole transport Ubiquitin-dependent protein catabolic process via the multivesicular body sorting pathway	Protein binding	Cytoplasm ESCRT III complex

Table 2. List of relevant genes obtained by YEASTRACT and grouped by motif. GO classification is described for each gene.

2.2.4 Search of transcription factors that recognize TFBSs inside motifs

The YEASTRACT database also makes publicly available up-to-date information on documented regulatory associations between TFs and DNA-binding sites in *S. cerevisiae*. Information in this database has been curated on precise tests of the associations between TFs and DNA-binding sites provided by experiments such as Chromatin ImmunoPrecipitation (ChIP), ChIP-on-chip and Electrophoretic Mobility Shift Assay (EMSA), that prove the direct binding of the TF to the target gene promoter region. Alternatively, the effect on target-gene expression of the site-directed mutation of the TF binding site in its promoter region was also considered by direct experimental evidence, which strongly suggests that the TF interacts with that specific target (Abdulrehman et al.,

2011). Analysis of TFs that bind to TFBSs from genes listed in Table 2 was performed for motifs 1 to 5. The TFs found to bind to motif 1 are shown in Table 3. The identified TFs are mainly involved in the control of the cell cycle and unfolded protein response, and to a lesser extent, in inter-organelle communication and energy metabolism. TFs that recognize motifs 2-5 also include Ash1p, Hac1p and Mot3p. Strikingly, the transcription factor Stb5p, an activator of multidrug resistance genes, binds to motifs 2, 4 and 5. Other TFs identified are also involved in the regulation of energy metabolism and cell cycle. It is important to point out that single base changes in the tested TFBSs dramatically increase the number of putative TFs that bind to them, suggesting that the identified TFs are not likely to be chosen randomly.

<table>
<tr><th>Transcription Factor</th><th>Consensus</th><th>Position</th><th>Strand</th><th colspan="2">Protein Info</th></tr>
<tr><td colspan="6">Target Sequence: Motif1_Subfamily I (size 92)</td></tr>
<tr><td>Ace2p, Swi5p</td><td>ACCAGC</td><td>-19</td><td>R</td><td>Transcription factor that activates expression of early G1-specific genes, localizes to daughter cell nuclei after cytokinesis and delays G1 progression in daughters, localization is regulated by phosphorylation; potential Cdc28p substrate</td><td>Transcription factor that activates transcription of genes expressed at the M/G1 phase boundary and in G1 phase; localization to the nucleus occurs during G1 and appears to be regulated by phosphorylation by Cdc28p kinase</td></tr>
<tr><td rowspan="2">Ash1p</td><td rowspan="2">YTGAT</td><td>-87, -48</td><td>F</td><td colspan="2" rowspan="2">Zinc-finger inhibitor of HO transcription; mRNA is localized and translated in the distal tip of anaphase cells, resulting in accumulation of Ash1p in daughter cell nuclei and inhibition of HO expression; potential Cdc28p substrate</td></tr>
<tr><td>-68</td><td>R</td></tr>
<tr><td>Hac1p</td><td>CCAGC</td><td>-20</td><td>R</td><td colspan="2">bZIP transcription factor (ATF/CREB1 homolog) that regulates the unfolded protein response, via UPRE binding, and membrane biogenesis; ER stress induced splicing pathway utilizing Ire1p, Trl1p and Ada5p facilitates efficient Hac1p synthesis</td></tr>
<tr><td rowspan="2">Mot3p</td><td>WAGGTA</td><td rowspan="2">-55</td><td rowspan="2">F</td><td colspan="2" rowspan="2">Nuclear transcription factor with two Cys2-His2 zinc fingers; involved in repression of a subset of hypoxic genes by Rox1p, repression of several DAN/TIR genes during aerobic growth, and repression of ergosterol biosynthetic genes</td></tr>
<tr><td>TAGGTA</td></tr>
</table>

Rtg1p, Rtg3p	GGTAC	-22	F	Transcription factor (bHLH) involved in interorganelle communication between mitochondria, peroxisomes, and nucleus	Basic helix-loop-helix-leucine zipper (bHLH/Zip) transcription factor that forms a complex with another bHLH/Zip protein, Rtg1p, to activate the retrograde (RTG) and TOR pathways
Target Sequence: Motif2_Subfamily I (size 90)					
Ash1p	YTGAT	-68	F	Zinc-finger inhibitor of HO transcription; mRNA is localized and translated in the distal tip of anaphase cells, resulting in accumulation of Ash1p in daughter cell nuclei and inhibition of HO expression; potential Cdc28p substrate	
Hac1p	CCAGC	-1	R	bZIP transcription factor (ATF/CREB1 homolog) that regulates the unfolded protein response, via UPRE binding, and membrane biogenesis; ER stressinduced splicing pathway utilizing Ire1p, Trl1p and Ada5p facilitates efficient Hac1p synthesis	
Stb5p	CGGNS	-58	F	Activator of multidrug resistance genes, forms a heterodimer with Pdr1p; contains a Zn(II)2Cys6 zinc finger domain that interacts with a PDRE (pleotropic drug resistance element) in vitro; binds Sin3p in a two-hybrid assay	
		-56	R		
Target Sequence: Motif3_Subfamily I (size 68)					
Mot3p	AAGGKA	-62	F	Described before	
Target Sequence: Motif4_Subfamily I (size 50)					
Stb5p	CGGNS	-25, -14	F	Described before	
		-18, -40	R		
Gsm1p	CGGNNNN NNNNCGG	-25	F	Putative zinc cluster protein of unknown function; proposed to be involved in the regulation of energy metabolism, based on patterns of expression and sequence analysis	
Target Sequence: Motif5_SubfamilyI (size 63)					
Gcr1p	CTTCC	-56	R	Transcriptional activator of genes involved in glycolysis; DNA-binding protein that interacts and functions with the transcriptional activator Gcr2p	

Mot3p	CAGGYA	-21	F	Described before
Stb5p	CGGNS	-35	F	Described before
Xbp1p	CTCGA	-51	F	Transcriptional repressor that binds to promoter sequences of the cyclin genes, CYS3, and SMF2; expression is induced by stress or starvation during mitosis, and late in meiosis; member of the Swi4p/Mbp1p family; potential Cdc28p substrate

Table 3. Transcription factors that recognize TFBSs in Motifs 1 - 5.

The recognized consensus sequence, relative position and bound strand is indicated. For each TF the protein information deposited in SGD and Yeastract is provided.

3. Discussion

This work was initiated as an attempt to understand and define the promoter structure of the 10 *lip* genes from the ligninolytic basidiomycete *P. chrysosporium,* assuming that the members of this family are co-regulated and have a common code for this particular biological function. The first encouraging hint was the discovery of common TFBS sequences which suggested a coordinated response to the various processes involved in lignin biodegradation. Furthermore, the presence of a common organization might permit the identification of additional genes in the *P. chrysosporium* genome that participate in lignin degradation, on the basis that they received similar regulatory "inputs".

Multiple alignment of all *lip* promoters yielded short homologous sequences that included experimentally validated TFBSs in other eukaryotic organisms, including yeast. These results were very encouraging. Hoping to find that similar promoters would present comparable physiological responses, transcriptional levels of *lip* genes of the fungus grown in C- and N-limited cultures were examined. However, no clear correlation between genomic organization and transcript levels was observed under these conditions. Analysis of the 10 promoters using multiple programs and databases only showed scattered and ambiguous (or degenerate) TFBSs and no clear structural organization emerged. The use of MEME software represented a breakthrough, since it allowed finding sequences that share a common (but hidden) property in conserved positions, which do not correspond to *a priori* experimentally determined TFBSs (called *Ab Initio*). MEME detected the five relevant and statistically significant motifs presented in this work. This is consistent with the group of six *lip* genes with a highly conserved gene and protein structure, which had been previously reported by Stewart and Cullen (1999). This finding suggested that the subfamily I *lip* genes derived from several duplication events of an ancestral gene.

The next task consisted in determining if there is a common biological function associated to each motif. For this, the sequence of each motif was analyzed in the YEASTRACT database which identified yeast genes which also contained any of the five motifs within their regulatory sequences. Indeed, one or more genes were found for each motif which contained curated and experimentally validated TFBSs. How do these genes relate to the biological process of lignin biodegradation? In order to answer this question, each motif was analyzed.

Motif 1 included a single gene associated to the cellular response to oxidative stress. During secretion of enzymes involved in the ligninolytic process, such as LiPs and MnPs, oxidative stress is a natural condition of *P. chrysosporium* and resistance to oxidative stress is probably an important function (Zacchi et al., 2000; Belinky et al., 2003; Jiang et al., 2005). To date there is no clear evidence in the literature on the mechanisms used by *P. chrysosporium* to tolerate the highly oxidative environment produced during lignin degradation. An orthologue of the yeast TRX2 gene, which encodes a cytoplasmic thioredoxin isoenzyme, could be involved in the protection of *P. chrysosporium* cells against oxidative and reductive stress. Motif 1 is also related to the secretion of vesicles, which is fully consistent with the manner in which these enzymes are carried into the extracellular medium.

Motif 2 is contained in several genes that do not share an obvious common function, although most of them are related to nitrogen metabolism, the Krebs cycle and ribosomal biogenesis. As is well known, LiPs are induced in response to low nitrogen and low carbon conditions, which suggests that the cell might be increasing protein synthesis, a necessary process for hyphal remodeling and growth.

Motif 3 is common to genes involved in cellular response to nitrogen and carbon metabolism, including gluconeogenesis. Some genes containing this motif are involved in the nitrogen cellular response to starvation and regulation of carbohydrate metabolic processes. There is a partial overlap of biological functions (though not of genes) with Motif 2, however, other interesting biological processes also seem to be involved: invasive growth in response to glucose limitation, which suggests remodeling of fungal cellular structures, such as cell wall assembly. It is known that during the ligninolytic process, *P. chrysosporium* apical tips of hyphae penetrate the wood through the tracheids and secrete ligninolytic enzymes. The yeast gene YDR477W | SNF1 contains motif 3 in its promoter and encodes an AMP-activated serine / threonine protein kinase, which is involved in signal transduction and found in a complex with proteins required for the transcription of glucose-repressed genes and involved in sporulation and peroxisome biogenesis. This gene might be related to stress tolerance regulation and gene expression under low carbon conditions, as would occur in secondary metabolism (ligninolysis), which is coupled to sporulation (structural remodeling of the fungus) and possibly, peroxisome biogenesis.

Motif 4 is present in the promoter of two yeast genes described as ion transporters which are of interest in relation to lignin biodegradation: genes YML121W | gtr1 and YML123C | PHO84 are involved in phosphate transport, which is essential for nucleic acids synthesis, and therefore also associated to cell cycle regulation, which in turn might be related to hyphal growth. YML121W | gtr1 encodes a cytoplasmic GTP binding protein and negative regulator of the Ran/Tc4 GTPase cycle; it is also a component of the GSE complex required for sorting of Gap1p and is involved in phosphate transport and telomeric silencing, similar to human Raga and Ragbir proteins. YML123C | PHO84 is a high-affinity inorganic phosphate (Pi) and low-affinity manganese transporter. The latter is relevant in the context ligninolysis since Mn^{+2} has a regulatory role in the formation of LiPs (Rabinovich et al., 2004). Transport of this ion is important for the expression and activity of all kinds of ligninolytic enzymes from *P. chrysosporium*. Motif 4 is the most distal motif identified in the *lip* gene promoters. Due to its location on the promoter, it is tempting to speculate that this motif might be involved in DNA looping.

Motif 5 appears to be related to mitosis, cell cycle, chromosome segregation and stress response. The two yeast genes with motif 5 in their promoters and selected with the greatest stringency by YEASTRACT, YFR028C | CDC14 and YGR098C | ESP1, are required for the regulation of mitotic exit. This correlates well with active cell division that occurs in hyphae.

Other promoters which contain Motif 5, such as those from genes YDL079C | MRK1 (a glycogen synthase kinase 3 (GSK-3)) homolog and YOR181W|LAS17 are stress responsive genes. Finally, gene YGR274C | TAF1 (which encodes a TFIID subunit and is involved in promoter binding and G1 / S progression) and gene YOR140W | SFL1 are RNA polymerase II regulators. These functions appear to be complementary to those associated to the other motifs.

How do these regulatory elements coordinate fungal metabolism in natural environments? It is well known that filamentous fungi grow by apical extension and lateral branching to form mycelial colonies (Richards et al., 2010). Because of key characteristics of hyphae, filamentous fungi can efficiently colonize and exploit the substratum on which they grow, e.g. wood (Weber, 2002). Fungal cells within a single mycelium are known to autolyse to provide nutrients to ensure growth (Zacchi et al., 2000), involving processes related to the remodeling of the mycelium. In fungi, vacuoles are very versatile organelles involved in protein turnover, cellular homeostasis, membrane trafficking, signaling and nutrition (Veses et al., 2008), as well as progression through cell cycle checkpoints (Richards et al., 2010). Networks of spherical and tubular vacuoles have been found in a range of filamentous fungi, including the wood rotting plant pathogen *Phanerochaete velutina* (Richards et al., 2010). Under LiP producing conditions, hyphal cells undergo a major loss of cellular ultrastructure, similar to that observed under oxidative stress (Zacchi et al., 2000). Therefore LiPs may be enzymes that are induced under conditions of oxidative stress (Rabinovich et al., 2004) and degrade lignin in order to access further carbon sources (Zacchi et al., 2000).

Taken together, many of the genes shown to contain any of these motifs have in common that they regulate genes of relevance associated to the biological processes that occur during lignin biodegradation. They include stress, mycelia remodeling which involves changes in lipid and carbohydrate metabolism, and mitosis, that lead to organellar /ultrastructural reorganization and changes related to the shift to secondary metabolism. In an analogous manner, transcription factors that apparently recognize these motifs, also bind TFBSs of genes involved in stress response and mitosis, among others (See Table 3).

4. Final remarks and conclusion

This work proposes an ordered and step by step approach for the analysis of the putative structure of eukaryotic promoters. To test this strategy, the *lip* gene family from the ligninolytic fungus *P. chrysosporium* was studied. The resulting analysis uncovered an organization of TFBSs into structural motifs that is not evident using standard software. The MEME software, which searches for signals that are shared by a group of sequences, was instrumental to detect these hidden elements. Each of the discovered motifs contains several TFBSs. One transcription factor may bind to various sites and hence it is speculated that the TFBS pairs group into clusters, which may be bound by the same transcription factor. Clusters with TATA-related and CAAT-related pairs have been reported (Ma et al., 2004). Also, several TATA-box related triples have been described in the literature (Ma et al., 2004). Each motif found in our analysis may represent this clustering of TFBSs and therefore may correspond to the basic functional unit of a promoter. The functional promoter may then be an organized sequence of motifs, as diagrammed in Figures 5 and 6. A simple sentence can be envisioned as an analogy of this regulatory structure: a sentence containing an instruction in any language corresponds to a meaningful sequence of words. The promoter represents this sentence and each motif corresponds to one of the words. In turn, as each word is composed by several syllables, each motif is built by combining several TFBSs. Just as syllables, which contain several letters, isolated TFBSs contain several nucleotides and may be present in more than one copy in a single word or appear in several different words within the same sentence, but often do not have functional meaning on their own.

In conclusion, this work proposes an ordered and step by step approach for the analysis of the putative structure of eukaryotic promoters. We devised a straightforward *in silico* strategy that permits the identification of promoter structure in a set of related eukaryotic genes. To test this strategy the *lip* gene family from the ligninolytic fungus *P. chrysosporium* was studied. The resulting analysis uncovered an organization of TFBSs into structural motifs (that are not evident using standard software) which are present in yeast genes and transcription factors involved in diverse processes related to the biological context in which ligninolysis is carried out. The structured motifs discovered in this study may represent a functional organization of regulatory sequences. A future challenge will be to test other gene families in order to determine if the proposed model is a general feature of eukaryotic systems.

5. Acknowledgments

We wish to thank Dr. Roberta Farrell, Waikato University, New Zealand, for revision of the manuscript.

6. References

Abdulrehman, D., Monteiro, P.T., Teixeira, M.C., Mira, N.P., Lourenço, A.B., dos Santos, S.C., Cabrito, T.R., Francisco, A.P., Madeira, S.C., Aires, R.S., Oliveira, A.L., Sá-Correia, I. & Freitas, A.T. (January 2011). YEASTRACT: providing a programmatic access to curated transcriptional regulatory associations in *Saccharomyces cerevisiae* through a web services interface. *Nucleic Acids Research*, Vol. 39, No. (Database issue, Epub 2010 Oct 23), pp. D136–D140.

Bailey, T.L. & Elkan, C. (1994). Fitting a mixture model by expectation maximization to discover motifs in biopolymers. *Proceedings of the Second International Conference on Intelligent Systems for Molecular Biology*, pp. 28-36, AAAI Press, Menlo Park, California, USA.

Bailey, T.L. & Gribskov, M. (1998). Combining evidence using p-values: application to sequence homology searches. *Bioinformatics*. Vol. 14, pp. 48-54.

Bailey, T.L., Bodén, M., Whitington, T., & Machanick, P. (2010). The value of position-specific priors in motif discovery using MEME. *BMC Bioinformatics*, Vol. 11 (April 2010), pp. 179.

Bailey, T.L., Williams, N. Misleh, C. & Li, W.W. (2006). MEME: discovering and analyzing DNA and protein sequence motifs. *Nucleic Acid Research*, Vol. 34, No. WebServer Issue, pp. W369-W373.

Belinky, P.A., Flikshtein, N., Lechenko, S., Gepstein, S., & Dosoretz, C.G. (2003). Reactive oxygen species and induction of lignin peroxidase in *Phanerochaete chrysosporium*. *Applied and Environmental Microbiology*. Vol. 69, No. 11 (November 2003), pp. 6500-6506.

Dashtban, M., Schraft, H. & Qin, W. (2009). Fungal conversion of lignocellulosic residues; opportunities & perspectives. *International Journal of Biological Sciences*, Vol. 5, No 6, pp. 578-595.

Dempster, A.P., Laird, N.M. & Rubin, D.B. (1977). Maximum Likelihood from Incomplete Data via the EM Algorithm. *Journal of the Royal Statistical Society, Series B (Methodological)*, Vol. 39, No. 1, pp. 1–38.

D'Souza, T.M., Merritt, C.S. & Reddy, C.A. (1999). Lignin-modifying enzymes of the white rot basidiomycete *Ganoderma lucidum*. *Applied and Environmental Microbiology*, Vol. 65, No. 12, pp. 5307-5313.

Farrell, R.L., Murtach, K.B., Tien, M, Mozuch, M.D., & Kirk, T.K. (1989). Physical and Enzymatic Properties of Lignin Peroxidase isoenzymes from *Phanerochaete chrysosporium*. *Enzyme Microbial Technology*, Vol. 11, No. 6, pp. 322-28.

Han, L., Garcia, H.G., Blumberg, S., Towles, K.B., Beausang, J.F. & Phillips, R. (2009). Concentration and length dependence of DNA looping in transcriptional regulation *Plos One*, Vol. 4, No 5., pp. e5621.

Haygood, R., Babbitt, C.C., Fedrigo, O. & Wray, G.A. (2010). Contrasts between adaptative coding and noncoding changes during human evolution. *Proceedings of the National Academy of Sciences.* Vol. 107, No. 17 (May 2010), pp. 7853-7857.

Hein, J. (1990) Unified approach to alignment and phylogenies. *Methods in Enzymology*, Vol. 183, pp. 626-645, Academic Press, Inc. San Diego, California, USA.

Howard, M.L. & Davidson, E.H. (2004). Cis-Regulatory control circuits in development. *Developmental Biology*, Vol. 271, No. 1, pp. 109-118.

Jiang, Q., Yan, Y.-H., Hu, G.-K. & Zhang, Y.-Z. (2005). Molecular cloning and characterization of a peroxiredoxin from *Phanerochaete chrysosporium. Cellular and Molecular Biology Letters*, Vol. 10, No. 4, pp. 659 - 668.

Jiang, X., Zeng, G., Huang, D., Chen, Y., Liu, F., Huang, G., Li, J., Xi, B. & Liu, H. (2006) Remediation of pentachlorophenol-contaminated soil by composting with immobilized *Phanerochaete chrysosporium*. *World Journal of Microbiology and Biotechnology*, Vol. 22, No 9, pp. 909-913.

Kirk, T.K. & Farrell R.L. (1987). Enzymatic "combustion": the microbial degradation of lignin. *Annual Review of Microbiology*, Vol. 41, pp. 465-505.

Kirk, T.K., Tien, M., Kersten, P., Kalyanaraman, B., Hammel, K., & Farrell, R.L. (1990). Lignin Peroxidase from Fungi *Phanerochaete chrysosporium*. *Methods of Enzymology*, Vol. 188, p. 159-171, Academic Press, Inc. San Diego, California, USA.

Lobos, S., Larraín, J., Salas, L., Cullen, D. & Vicuña, R. (1994). Isoenzymes of manganese-dependent peroxidase and laccase produced by the lignin degrading basidiomycete *Ceriporiopsis subvermispora*. *Microbiology*, Vol. 140, pp. 2691–2698.

Ma, X.-T., Qian, M.-P. & Tang H.-X. (2004). Predicting Polymerase II Core Promoters by Cooperating Transcription Factor Binding Sites in Eukaryotic Genes. *Acta Biochimica et Biophysica Sinica*, Vol. 36, No. 4, pp. 250–258.

Mariño-Ramírez, L., Spouge, J.L., Kanga, G.C. & Landsman, D. (2004). Statistical analysis of over-represented words in human promoter sequences. *Nucleic Acids Research*, Vol. 32, No. 3, pp. 949-958.

Martínez, D., Larrondo, L.F., Sollewijn, G., Huang, K., Helfenbein, K.G., Ramaiya, P., Detter, J.C., Larimer, F., Coutinho, P.M., Henrissat, B., Berka, R., Cullen, D., & Rokhsar, D.

(2004). Genome sequence of the lignocelluloses degrading fungus *Phanerochaete chrysosporium* strain RP78. *Nature Biotechnology*, Vol. 22, pp. 695-700.

Martínez, D., Challacombe, J., Morgenstern, I., Hibbett, D., Schmoll, M., Kubicek, C.P., Ferreira, P., Ruiz-Duenase, F.J., Martinez, A.T., Kersten, P., Hammel, K., Wymelenbergg, A.V., Gaskell, J., Lindquist, E., Sabat, G., Splinter, S., BonDuranti, Larrondo, L.F., Canessa, P., Vicuña, R., Yadav, J., Doddapaneni, H., Subramanian, V., Pisabarro, A.G., Lavín, J.L., Oguiza, J.A., Master, E., Henrissat, B., Coutinho, P.M., Harris, P., Magnuson, J.K., Baker, S.E., Bruno, K., Kenealy, W., Hoegger, P.J., Kües, U., Ramaiya, P., Lucas, S., Salamov, A., Shapiro, H., Tu, H., Chee, C.L., Misra, M., Xie, G., Teter, S., Yaver, D., James, T., Mokrejs, M., Pospisek, I.V. Grigoriev, T. Brettin, D. Rokhsar, Berka, R. & Cullen, D. (2009). Genome, transcriptome, and secretome analysis of wood decay fungus *Postia placenta* supports unique mechanisms of lignocellulose conversion. *Proceedings of the National Academy of Sciences*, Vol. 106, No. 6, pp. 1954–1959.

Matthews, K.S. (1992). DNA looping. *Microbiology Review*, Vol. 56, No. 1, pp. 123-136.

Monteiro, P.T., Mendes, N.D., Teixeira, M.C., d'Orey, S., Tenreiro, S., Mira, N.P., Pais, H., Francisc, A.P., Carvalho, A.M., Lourenço, A.B., Sá-Correia, I., Oliveira, A.L. &Freitas, A.T. (2008). YEASTRACT-DISCOVERER: new tools to improve the analysis of transcriptional regulatory associations in *Saccharomyces cerevisiae*. *Nucleic Acids Research*, Vol. 36, No. 36 Database issue (January 2008), pp. D132-D136.

Patikoglou, G.A., Kim, J.L., Sun, L., Yang, S.H., Kodadek, T. & Burley, S.K. (1999). TATA element recognition by the TATA box-binding protein has been conserved throughout evolution. *Genes and Development*, Vol. 13, (Dec 1999), No. 24, pp. 3217-3230.

Rabinovich, M.L., Bolobova, A.V. & Vasil'chenko L.G. (2004). Fungal decomposition of natural aromatic structures and xenobiotics: a review. *Applied Biochemistry and Microbiology*, Vol. 40, No. 1, pp. 1-17.

Rajakumar, S., Gaskell, J., Cullen, D, Lobos, S, Karahanian, E. & Vicuña, R. (1996). Lip-Like genes in *Phanerochaete sordida* and *Ceriporiopsis subvermispora*, white rot fungi with no detectable lignin peroxidase activity. *Applied and Environmental Microbiology*, Vol. 62, No. 7 (July 1996), pp. 2660–2663.

Richards, A., Veses, V. & Gow, N.A.R. (2010). Vacuole dynamics in fungi. *Fungal Biology Reviews*, Vol. 24, pp. 93-105.

Rüttimann-Johnson, C., Cullen, D. & Lamar, R. (1994). Manganese peroxidases of the white rot fungus *Phanerochaete sordida*. *Applied and Environmental Microbiology*, Vol. 60, pp. 599–605.

Sánchez, C. (2009). Lignocellulosic residues: Biodegradation and bioconversion by fungi. *Biotechnology Advances*, Vol. 27, No. 2 (March-April 2009), pp. 185-194.

Sato, S., Liu, F., Koc, H. & Tien, M. (2007). Expression analysis of extracellular proteins from *Phanerochaete chrysosporium* grown on different liquid and solid substrates. *Microbiology*, Vol. 153, No. Pt 9 (September 2007), pp. 3023-3033.

Schleif, R. (1992). DNA looping. *Annual Review of Biochemistry*, Vol. 61, pp. 199-223.

Shi, J., Chinn, M.S. & Sharma-Shivappa, R.R. (2008). Microbial pretreatment of cotton stalks by solid state cultivation of *Phanerochaete chrysosporium*. *Bioresources Technology*, Vol. 99, No. 14 (September 2008), pp 6556-6564.

Shi, J., Sharma-Shivappa, R.R. & Chinn, M.S. (2009). Microbial pretreatment of cotton stalks by submerged cultivation of *Phanerochaete chrysosporium*. *Bioresources Technology*, Vol. 100, No. 19 (October 2009), pp. 4388-4395.

Singh, D. & Chen, S. (2008). The white-rot fungus *Phanerochaete chrysosporium*: conditions for the production of lignin-degrading enzymes. *Applied Microbiology and Biotechnology*, Vol. 81, No. 3 (December 2008), pp. 399-417.

Singh, P., Sulaiman, O., Hashim, R., Rupani, P.F. & Peng, L.C. (2010). Biopulping of lignocellulosic material using different fungal species. *Reviews in Environmental Science and Biotechnology*, Vol. 9, pp. 141-151.

Smith, T.F. & Waterman, M.S. (1981). Identification of common molecular subsequences. *Journal of Molecular Biology*, Vol. 147, No. 1, pp. 195-197.

Stepanova, M., Tiazhelova, T., Skoblov, M. & Baranova, A. (2005). A comparative analysis of relative occurrence of transcription factor binding sites in vertebrate genomes and gene promoter areas *Bioinformatics*, Vol. *21*, No. 9 (May 2005), pp. 1789–1796.

Stewart, P. & Cullen, D. (1999). Organization and Differential Regulation of a cluster of lignin Peroxidase genes from *Phanerochaete chrysosporium*. *Journal of Bacteriology*, Vol. 181, No. 11 (June 1999), pp. 3427-3432.

Thompson, J.D., Higgins, D.G. & Gibson, T.J. (1994). CLUSTAL W: improving the sensitivity of progressive multiple sequence alignment through sequence weighting, position-specific gap penalties and weight matrix. *Nucleic Acids Research*, Vol. 22, No. 22 (November 1994), pp. 4673-4680.

Veses, V., Richards, A. & Gow, N.A.R. (2008). Vacuoles and fungal biology. *Current Opinion in Microbiology*, Vol. 11, pp. 503-510.

Weber, R.W.S. (2002). Vacuoles and the fungal lifestyle. *Micologist*, Vol. 16. Part 1, pp. 10-20.

Wei, W. & Yu. X.-D. (2007). Comparative analysis of regulatory motif discovery tools for Transcription Factor Binding Sites. *Genomics Proteomics Bioinformatics*, Vol. 5, No. 2 (May 2007), pp. 131-142.

Wittkopp, P.J. (2006). Evolution of *cis*-regulatory sequence and function in Diptera. *Heredity*, Vol. 97, No. 3 (July 2007), pp. 139-147.

Wu, J.M. & Zhang, Y.Z. (2010). Gene Expression in Secondary Metabolism and Metabolic Switching Phase of *Phanerochaete chrysosporium*. *Applied and Biochemical Biotechnology*, Vol. 162, No. 7 (November 2010), pp. 1961-1977.

Wymelenberg, A.V., Gaskell, J., Mozuch, M., Kersten, P., Sabat, G., Martinez, D. & Cullen, D. (2009). Transcriptome and Secretome Analyses of *Phanerochaete chrysosporium* Reveal Complex Patterns of Gene Expression. *Applied and Environmental Microbiology*, Vol. 75, No. 12 (June 2009), pp. 4058–4068.

Zacchi, L. Morris, I. & Harvey, P.J. (2000). Disordered ultrastructure in lignin-peroxidase-secreting hyphae of the white-rot fungus. *Phanerochaete chrysosporium*. *Microbiology*, Vol.146, pp. 759-765.

5

Prediction and Experimental Detection of Structural and Functional Motifs in Intrinsically Unfolded Proteins

Cesira de Chiara and Annalisa Pastore
MRC National Institute for Medical Research, The Ridgeway, London, UK

1. Introduction

Intrinsically unstructured proteins (IUPs) or proteins with intrinsically unstructured regions (IURs) have quickly gained increasing interest within the biological community because of their significant presence in the human genome and their potential links to major pathologies such as cancer, neurodegeneration and diabetes (Tompa, 2005; Tompa & Fuxreiter, 2008). The terms IUPs and IURs designate proteins or protein regions intrinsically devoid of a well defined tertiary structure. The concept was introduced a few years ago in the scientific literature as a brand new idea, which would represent a family of proteins thought to have been previously ignored or unappreciated (Dunker et al., 2001; Wright & Dyson, 1999). However, as in many other examples in Science, the concept of IUPs is far from being new. In the 70s, it was universally accepted that what were known as 'biologically active peptides' had no intrinsic structure, often being too short to have a proper hydrophobic core. Peptides would/could however fold in a definite conformation upon interaction with a partner/receptor, thus having all the features of modern IUPs (Boesch et al., 1978). Hormones and opioid peptides are two among several of the best studied examples.

The concept of intrinsic disorder and/or flexibility has now been extended to proteins and has deeply transformed our perception of the importance of protein dynamics as opposed to the static picture introduced by years of crystallographic studies. Even more important is the fact that accepting the existence of IUPs proposes a unique paradigm in which function can be directly linked to structural disorder rather than to a defined structure.

IUPs have been classified in two broad categories. In the first family, IUP's function is achieved through binding to one or several partner molecule(s) in a structurally adaptive process, which enables an exceptional plasticity in cellular responses. These proteins do not form a structure by themselves and are functionally inactive in the absence of a partner, but structure can be induced upon recognition of another molecule. When bound to a substrate, they are able to acquire a structure and become rigid, according to an induced-fit mechanism or to what has been recently generalized in the concept of 'conformational fuzziness' (Wright & Dyson, 2009). Macromolecular association rates have in fact been demonstrated to be highly enhanced by a relatively non-specific association enabled by flexible recognition segments. Molecular recognition occurring in this way has been described according to the 'fly-casting' (Shoemaker et al., 2000) or 'protein fishing' (Evans & Owen, 2002) mechanisms. Examples of proteins belonging to this family are those bearing RNA binding motifs which acquire a structure only upon interaction to RNA.

In the second family, IUPs work as entropic chains exploiting their ability of fluctuating over ensembles of structural states with similar conformational energies. In this way they either generate force against structural changes or influence the orientation/localization of attached domains (Dunker et al., 2002). According to these properties, they are active in their unstructured form and play the role of flexible linkers necessary to allow other portions of the protein to move like 'a dog on a leash' and ultimately to interact with other partners. A classical example of such a case is the IUR, called PEVK, of the muscle protein titin (Labeit & Kolmerer, 1995). This region confers some of the passive elastic properties of the titin filament, providing the stiffness required in muscle contraction to keep the sarcomere in register (Greaser et al., 2000).

These unique features are exploited in many biological processes thus explaining the multiplicity of different functions in which IUPs are involved (Dunker et al., 2002). Protein disorder prevails for instance in signaling, regulatory and cancer-associated proteins. The functional importance of protein disorder is also underlined by its dominant presence in proteins associated with signal transduction, cell-cycle regulation, gene expression and chaperone activities (Dunker et al., 2002; Iakoucheva et al., 2002; Tompa & Csermely, 2004; Uversky, 2002; Ward et al., 2004b). Because of their susceptibility to degradation, IUPs have also been linked to the ubiquitin (Ub)/proteasome pathway (Csizmok et al., 2007).

In this chapter, we discuss the problems related to the prediction, production and characterization of IUPs/IURs. We take as a representative case study ataxin-1, a human protein of biological and clinical interest that is related to the neurodegenerative disease spinocerebellar ataxia of type-1. Using ataxin-1, we retrace how the application of different bioinformatics tools has contributed to shed light on the structure and functions of the protein since its first identification. We also provide an update on the physico-chemical methods used to translate the sequence information into structural and functional models of the protein and its interactome. This hands-on example might provide a valuable paradigm of how correct identification of linear motifs and IURs can provide the key for understanding protein function.

2. Prediction methods of IUPs

While it has become increasingly easy to appreciate the importance of IUPs/IURs, their prediction and experimental characterization remain somewhat problematic. Since Romero et al. (Romero et al., 1997) indicated for the first time that lack of a defined protein tertiary structure is predictable on the sole basis of the primary sequence, several different methods have been developed that enable prediction (reviewed in Radivojac P et al. (Radivojac et al., 2007) (**Table 1**). They are based on different definitions of IURs and detect different indicators such as hydrophobicity, sequence composition, secondary structure content, etc.

These programs, however, are not always entirely reliable. Most weaknesses arise from the intrinsic difference between the conceptual and operational definition of IUP/IUR. As mentioned above, conceptually, there are two classes of IUP/IUR. The first does not form a structure by itself, but this can be induced by a partner. Proteins belonging to the second class can perform their function in three ways: (1) through the newly acquired structure, (2) by inducing structural changes in their partners and modulating the function of these partners, or (3) through formation of protein complexes with partners. This class of IUP/IUR does not function in its unstructured form, i.e., the unstructured form is inactive. The second class performs its cellular function without forming a structure: the unstructured form is the functional one.

Software	Prediction	Based on
Disopred2 (http://bioninf.cs.ucl.ac.uk/disopred) (Ward et al., 2004b)	Regions lacking ordered secondary structure	Cascaded support vector machine classifiers trained on PSI-BLAST profiles
PONDR (http://www.pondr.com/) (Radivojac et al., 2003)	Non rigid regions from random coils to partially unstructured regions and molten globules	Local aminoacid composition, flexibility, hydropathy, etc
GlobPlot (http://globplot.embl.de/) (Linding et al., 2003b)	Regions with high propensity for globularity based on propensity for secondary structure and random coil	Russell/Linding scale of disorder
DisEMBL (http://dis.embl.de/) (Linding et al., 2003a)	LOOPS (region without secondary structure) HOT LOOPS (loops with high mobility) REMARK465 (regions of crystal structures lacking of electron density)	Neural networks trained on crystal structures
IUPred (http://iupred.enzim.hu/) (Dosztanyi et al., 2005a, 2005b)	Regions devoid of well-defined 3D structure under native conditions	Local aminoacid composition used to estimate inter-residue interaction energy

Table 1. List of predictors of protein disorder used for ataxin-1 in this study. The table, which illustrates the features of the different methods, is adapted from Ferron et al. (Ferron et al., 2006).

The operational definition used by bioinformatic software is typically based on the likelihood of the protein/peptide forming a structure under certain (often poorly defined) cellular environments, with little or no information on how the protein/peptide functions or what partner the IUP/IPR may have. Intrinsically unstructured proteins are for instance characterized by a low content of bulky hydrophobic amino acids and a high percentage of polar and charged amino acids. As a consequence, they do not contain enough residues to build a hydrophobic core that is typical of stably folded globular proteins. Another symptomatic indication of an IUP is the presence of low complexity motifs, i.e. sequences with over-representation of just a few residues (e.g. polyglutamine stretches, arginine-glycine (RG) repeat observed in RNA binding proteins or arginine-serine (RS) repeats observed in splicing factors). Of course, while low complexity sequences are a strong indication of disorder, the reverse is not necessarily true, that is, not all disordered proteins have low complexity sequences.

We can then expect that as these predictions become more robust the more we shall be able to distinguish between the different IUPs/IURs and have functional clues. For the time being, this software should be handled with care. For instance it is quite important to use more than one approach on the same sequence and compare the resulting scores.

3. Detection of linear functional motifs

It is difficult to underestimate the importance of the concept of structural motifs or modules in the world of globular proteins (Konagurthu & Lesk, 2010; Schaeffer & Daggett, 2010). A similar role is taken now by linear motifs in the world of IUPs. Eukaryotic linear motifs (ELMs) are short stretches of eukaryotic protein sequences (typically 3-10 amino acids long) to which a molecular function is associated. These segments provide regulatory functions independently of protein tertiary structure. While also potentially present in stably folded proteins, they acquire a particular importance in IUPs because in these they cannot be shielded by structure. It is in fact found experimentally that short functional sites, which are frequently involved in regulatory processes, must reside in non-structured or non-globular regions or, when within globular domains, in flexible highly exposed loops (Gibson, 2009). Examples of linear motifs are phosphorylation sites, nuclear localization signals or signalling sequences. A useful tool for the prediction of linear motifs is the ELM database (http://elm.eu.org/) developed from a collaborative effort between EMBL and University of Rome (Diella et al., 2008; Puntervoll et al., 2003). A sequence of interest can be screened against this database to quickly suggest the position of ELMs.

4. When can we be sure that a protein is a IUP?

Many of us have been confronted with natural or recombinant proteins that, after purification, result in being unfolded even when they are expressed in a soluble form. Are these bona fide IUPs? It is some times difficult to distinguish between IUPs and orphan complex proteins, yet there is a profound difference between the two families. IUPs have no intrinsic capability to adopt a definite structure at least in the absence of a partner. Their conformational state is also independent of the way they have been purified. On the contrary, Orphan complex proteins are able to fold but their tertiary fold is only marginally stable even though not being intrinsically devoid of a 3D structure (Sjekloća et al. 2011). This is often due to the absence of a partner which stabilizes their fold. These proteins might be produced as unfolded proteins, as folded/unfolded mixed populations or as species highly prone to aggregation. Their misbehaviour can in principle be reduced or neutralized by finding more suitable conditions or less drastic purification protocols (e.g. concentration is often a problematic step). The only guidance to distinguish among the two families may be sequence analysis coupled to comparison with other homologous. If the motif is observed in other proteins where it adopts a stable fold, it is difficult to believe that it is intrinsically unfolded. In such a case, we suggest putting more care in protein production and attempting to identify the partners able to provide stabilization. A way to solve the ambiguity when the protein of interest is purified as an unfolded monomer can be to register an NMR spectrum directly on the cell lysate, if proper overexpression can be achieved. This circumvents the purification step and 'shows' the structural properties of the protein independently of human intervention.

5. Experimental methods to study IUPs

How can we study the structure of intrinsically unstructured molecules? The statement sounds like an oxymoron in which the two terms contradict each other. The problem closely resembles the difficulties encountered in the study of denatured states of ordered proteins. It goes without saying that IUPs cannot be crystallized in isolation because of their flexibility. Solution studies are therefore more appropriate to characterize their structural states. The ultimate goal of these studies cannot, however, be that of obtaining *the* structure but rather to describe the protein as an ensemble of rapidly interconverting alternative structures characterized by differing backbone torsion angles.

Among the biophysical methods able to provide structural and dynamic information, two techniques are probably the ones that have given more interesting results over the last few years: Nuclear magnetic resonance (NMR) and Small Angle X-ray Scattering (SAXS). Despite the disadvantages of the very poor chemical shift dispersion (particularly for proton and aliphatic carbon resonances) because of conformational averaging, NMR is particularly powerful thanks to its ability of measuring different independent observables. These include secondary chemical shifts, residual dipolar couplings (RDC), hydrogen bonds, torsional angles and long-range NOE upon spin labelling (for an exhaustive review see Mittag & Forman-Kay, 2007). Among them, detection of residual secondary chemical shifts is probably the simplest qualitative way to detect complete or partial local disorder of a chain. Most of the other NMR observables can be exploited in a more quantitative way: since the early implementation of the ensemble-averaged nOe distance restraints (Bonvin & Brunger, 1996), a variety of restraining algorithms, including simultaneous time and ensemble averaging (Fennen et al., 1995), have been developed and have been used to describe native, transition, intermediate, and unfolded states (Clore & Schwieters, 2004a, 2004b; Kuszewski et al., 1996; Vendruscolo & Paci, 2003; Vendruscolo & Dobson, 2005). A modern approach that takes into account protein flexibility is that of imposing penalizing forces if the calculated average distances at a given time across an ensemble of simulated molecules (the 'replica ensemble') do not match the experimental ones. It is interesting to note that RDCs provide a particularly powerful way to assess protein structures using an absolute reference system despite the original scepticism towards applying these measurements for the treatment of IUPs and IURs which pose fundamental problems in the way structural averaging should be used. To address the problem, new methods based on RDC measurements have been developed to provide detailed information on protein dynamics also in cases of conformational averaging, for instance based on analytical deconvolution, Gaussian axial fluctuation methods and restrained molecular dynamics simulations. Most of these methods have, however, been used to assess the dynamics of relatively small amplitude methods. Their application to flexible and conformationally interconverting molecules such as IUPs and IURs remains to be fully established.

SAXS, by which it is possible to achieve the measurement of molecular dimensions and a description of the overall shape of the ensemble, is, in many ways, complementary to NMR. A key advancement to quantitatively characterizing flexible proteins in solution by SAXS was achieved with the implementation of the approach known as ensemble optimization method (EOM) (Bernado et al., 2007). In this approach, flexibility is taken into account by postulating the coexistence of different conformational states for the protein contributing to the experimental scattering pattern. The different conformers can then be selected by genetic algorithms from a pool containing a large number of randomly generated models covering the protein configurational space. The EOM selected models are then analysed by quantitative statistical criteria also developed to determine the optimal number of conformers necessary to represent the ensemble. When possible, the quality of the analysis is increased by simultaneous fitting of multiple scattering patterns from deletion mutants, a procedure which is somewhat equivalent to improving the quality of sequence alignment introducing information about sequence homologues. The EOM protocol has now been validated by applying it to the study of several examples of completely or partially unfolded proteins and on multidomain proteins interconnected by flexible linkers and has shown to be a robust and helpful approach to the study of IUPs.

6. Ataxin-1 as a case study

To discuss the problems related to the prediction and characterization of IURs we chose as a representative case study ataxin-1, a human protein of biological and clinical interest that is related to the neurodegenerative disease spinocerebellar ataxia of type-1 (SCA1). This hereditary pathology is dominant and a member of a small family of neurodegenerative diseases linked to protein aggregation and misfolding, all caused by anomalous expansion of polyglutamine (polyQ) tracts in the gene coding region (Orr & Zoghbi, 2007) (**Figure 1**).

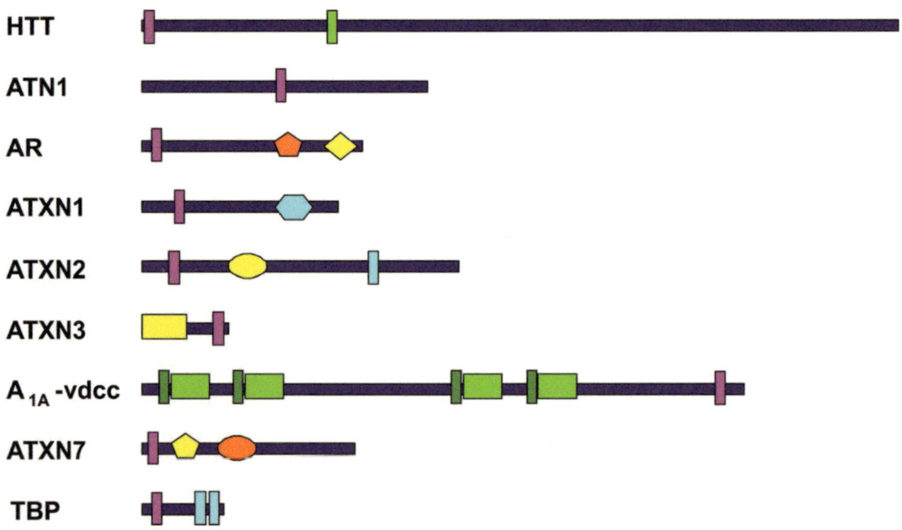

Fig. 1. Schematic representation of the protein members of the polyQ disease family. The sequence predicted to be intrinsically unfolded according to predictor DisEMBL (Radivojac et al., 2003) is indicated by a blue line. The position of polyQ tracts is indicated with magenta boxes. The position of other structural and functional domains, as reported by SMART (Schultz et al., 1998) and Pfam (Finn et al., 2010) databases, is indicated with geometric symbols of different colours. HTT: Huntingtin; ATN1: Atrophin 1; AR: Androgen receptor; ATXN1: Ataxin-1; ATXN2: Ataxin-2; ATXN3: Ataxin-3; A_{1A}-vdcc: A_{1A}-voltage dependent calcium channel; ATXN7: Ataxin-7; TBP: TATA-box binding protein.

Other members of the polyQ pathology family are Huntington's chorea, Machado-Joseph disease and other spinocerebellar ataxias. Elongation of polyQ tracts is the result of a gene polymorphism where unstable consecutive CAG triplets may become expanded during DNA replication. The mutated proteins form intracellular aggregates which are thought to be toxic for the cell and lead to cell death according to a gain rather than loss of function mechanism. While generally accepted that polyQ expansion is the leading factor in triggering pathology, it has become increasingly clear that protein context is an important element that modulates protein behaviour and pathology. Despite best efforts, all the polyQ pathologies are currently incurable. A way to develop a specific treatment is through the identification of the function(s) of the native proteins. In this endeavour, the support of bioinformatics analysis of the protein sequence becomes essential to predict structure and a great help in suggesting a function for these proteins. The sequences of the polyQ disease families are very different in length and position of the polyQ tract. They are also not homologous and share only a rather loose common feature: they all seem to contain large IURs which are sometimes interrupted by readily identifiable globular domains. As such, they constitute an excellent example to discuss the problems inherent in detecting and studying IUPs/IURs.

In the following sections we shall discuss a detailed analysis of the ataxin-1 primary sequence and how this study has suggested working hypotheses which could then be tested experimentally using structural and cellular approaches. We discuss the identification of two major potential IURs in ataxin-1 which contain short linear motifs important for phosphorylation, aggregation and protein-protein interactions that are directly related to pathogenesis (Chen et al., 2003; de Chiara et al., 2009; Emamian et al., 2003; Jorgensen et al., 2009; Klement et al., 1998). We pay particular attention to the prediction and characterization of the structural and functional features of the different protein regions and to the identification of key ELMs of crucial importance both for the normal and anomalous behaviour of the protein.

6.1 Ataxin-1 domain architecture

Ataxin-1 is a ca. 98 kDa protein which is well conserved through vertebrates, ubiquitously expressed and mainly localised in the cell nucleus (Klement et al., 1998). As with all the other members of the polyQ protein family, which it shares a common pathogenic mechanism with, ataxin-1 was originally identified by gene mapping (Banfi et al., 1994). In its non-expanded form, human ataxin-1 is a protein of 816 amino acids although expansion can significantly increase its length (**Figure 2**). Different predictions of IURs suggest the presence of long unstructured regions (**Figure 3**).

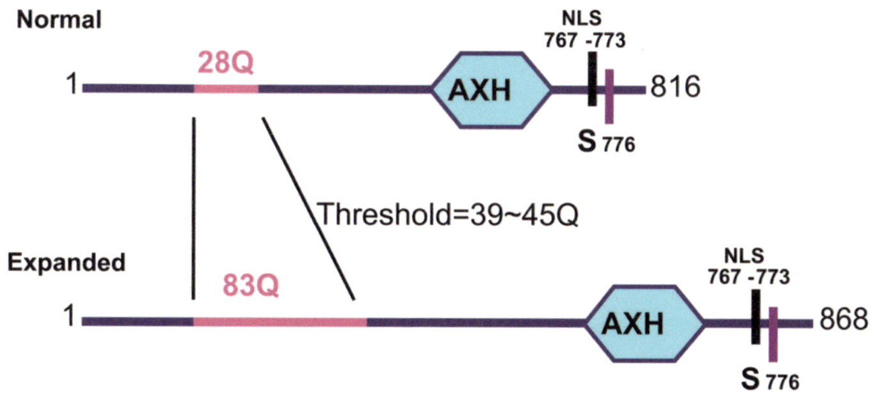

Fig. 2. Architecture of non-expanded and expanded ataxin-1. The polyQ tract is shown in magenta and the AXH domain in cyan. The positions of the nuclear localization signal (NLS) and a phosphorylation site important for protein interactions and for pathology are also indicated (de Chiara et al., 2009).

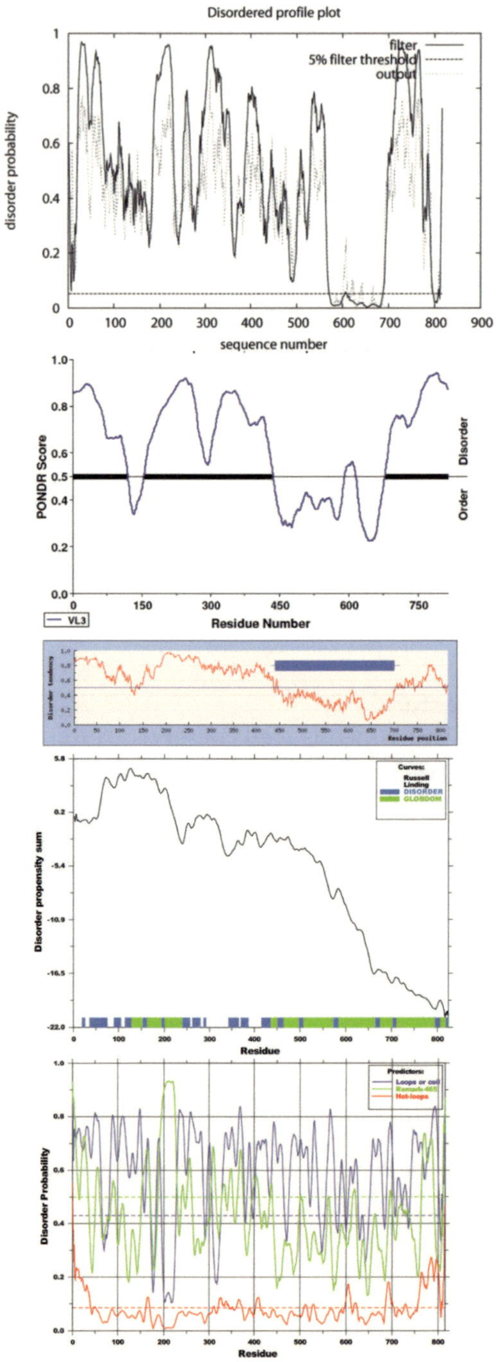

Fig. 3. IUR regions of ataxin-1 according to different predictors. From top to bottom: Disopred2 (Ward et al., 2004a; Ward et al., 2004b), PONDR (Radivojac et al., 2003), IUPRED (Dosztanyi et al., 2005a, 2005b), GlobPlot (Linding et al., 2003b) and DisEMBL (Linding et al., 2003a).

6.1.1 The N-terminal IUR

6.1.1.1 The N-terminal IUR according to different IUR predictors

In addition to the low complexity polyQ tract, there is a long region potentially unfolded from the N-terminus up to the AXH domain (567-689) which seems to be the only region of the protein that is autonomously folded (**Figure 2**). According to GlobPlot (Linding et al., 2003b), for instance, the region of ataxin-1 from the N-terminus to the beginning of the AXH domain is predicted as a series of multiple disorder regions. Among these are short low-complexity sequences (47-64, 88-99, 154-169, 366-377) as defined by SMART (Letunic et al., 2009; Schultz et al., 1998). Disopred2 (Ward et al., 2004a; Ward et al., 2004b), PONDR (Radivojac et al., 2003), DisEMBL (Linding et al., 2003a) and IUPRED (Dosztanyi et al., 2005a, 2005b) also concordantly predict this region as almost entirely disordered (**Figure 3**).

6.1.1.2 The polyQ tract

The N-terminus of the human protein is characterized by the presence of a highly polymorphic almost uninterrupted polyQ stretch which ranges from 4 to ~39 Qs in normal population and is expanded to ~40-83 Qs in SCA1-affected individuals (Zoghbi & Orr, 2008). Pathology typically develops when the repeat length exceeds a threshold of 35-45 glutamines (Genis et al., 1995; Jayaraman et al., 2009; Orr et al., 1993). Expansion of this region is a feature not shared among other species suggesting an evolutionary gain associated only to humans. The polyQ tract of ataxin-1 starts at residue 197. Indeed, the structure of polyQ stretches in solution has been shown experimentally both by CD and NMR spectroscopy to be a random coil when in a non-aggregated form (Masino et al., 2002). This is at variance with predictions by SMART which propose a helical coiled-coil region for the same region (amino acids 193-230). The discrepancy should anyway be ascribed by a bias in SMART for poly-amino acids. Interestingly, we now know that expansion of the polyQ tract in ataxin-1 is a condition necessary but not sufficient for triggering disease: two other motifs, a nuclear localization signal (NLS) and phosphorylation of S776, both located at the C-terminus and discussed in a session below, have been proved to also be required. As for the other polyQ proteins the native function of the polyQ tracts is unknown, although their presence has mostly been detected in proteins associated with transcriptional regulation activity. Indeed, the transcriptional regulator poly-Q binding protein-1 (PQBP1) has been found to bind ataxin-1 in a polyQ length-dependent manner, suggesting that PQBP1 and mutant ataxin-1 may act cooperatively to repress transcription and induce cell death (Okazawa et al., 2002). Direct evidence to support this hypothesis is now necessary.

6.1.1.3 Prediction of significant ELMs in the N-terminal IUR

Although ataxin-1 has been reported as a protein shuttling in and out the nucleus and the cytoplasm, there is a large body of evidence showing that the protein is predominantly located in the nucleus. Restricting the search for candidate short linear motifs and post-translational modifications within the N-terminal IUR to those significant for the nuclear localization, several ELMs have been predicted by using the ELM resource (Diella et al., 2008; Puntervoll et al., 2003) (**Table 2**).

A plethora of potential phosphorylation sites are predicted by the ELM resource in the N-terminal IUR of ataxin-1, among which only Ser239 (Vierra-Green et al., 2005) and Ser254 have been experimentally verified (Dephoure et al., 2008). Phosphorylation of these two serine residues supports the prediction by ELM of two candidate Class IV WW domain interaction motifs present in the regions 236-241 and 251-256 which mediate phosphorylation-dependent interactions. In addition to the WW domain motifs, other

ELM	Sequence	Position	Description	Experimentally Verified
LIG_14-3-3_1	RRWSAP	773-778	Mode 1 interacting phospho-motif for 14-3-3 proteins	Yes
LIG_14-3-3_3	KAPTLP RTASPP KAESSR HSASEP	29-34 260-265 309-314 439-444	Interacting phospho-motif for 14-3-3 proteins not matching Mode 1 and Mode 2	
LIG_FHA_1	EGTAWLP PVTSAVA PHTLTLG PGTQPLL FVTTALP	41-47 154-160 279-285 486-492 523-529	Phosphothreonine motif binding FHA with preference for large aliphatic at pT+3	
LIG_FHA_2	AGTSVEL LSTGLDY QATHREA PSTLNDK GSTDMEA FLTKIEP	63-69 81-87 400-406 408-414 496-502 758-764	Phosphothreonine motif binding FHA with preference for acidic aminoacid at pT+3	
LIG_WW_4	TPGSPP ISSSPQ	236-241 251-256	Phosphopeptide motif interacting with Class IV WW domain	
MOD_PKA_1	RRWSAPE	773-779	PKA-type AGC kinase phosphorylation site	Yes[a]
MOD_ProDKin_1	TPGSPPP ISSSPQN	236-242 251-257	MAPK phosphorylation site in higher eucaryotes	Yes[b] Yes[b]
MOD_CK2	RRWSAPE	773-779	CK2 phosphorylation site	
MOD_GSK3_1	ISSSPQNT RRWSAPES	251-258 773-780	GSK3 phosphorylation recognition site	Yes[b]
LIG_ULM_U2AF65_1	RKRRWS	771-776	Pattern in ULMs of SF1 and SAP155 which bind to the UHM of U2AF65	Yes
LIG_USP7_1	PATSR AGTSV PVTSA AVASA AESSR PYESR PSPSD ASPST PVGST PKPSL	20-24 63-67 154-158 158-162 310-314 356-360 366-370 406-410 494-498 795-799	Variant of the USP7 NTD domain based on the MDM2 and P53 interactions	
MOD_SUMO	PKSE LKTE	529-532 594-597	Motif recognized for modification by SUMO	
TRG_NLS_MonoCore_2	TRKRRW	770-775	Monopartite variant of the classical basically charged NLS. Core version	Yes

Table 2. Prediction of relevant functional linear motifs in the N- and C-terminus of human ataxin-1 by the ELM database (http://elm.eu.org/) (Diella et al., 2008; Puntervoll et al., 2003). Among all the predicted phosphorylation sites only the ones which have been

confirmed in vivo and are supportive of the prediction of other phosphopeptide motifs have been included. [a]Phosphorylation site and kinase experimentally confirmed in cerebellum (Jorgensen et al., 2009). [b]Phosphorylation site confirmed, kinase not identified (Dephoure et al., 2008; Vierra-Green et al., 2005)

putative phosphorylation-dependent protein-protein interaction motifs are predicted in the N-terminus of ataxin-1. Among these are several forkhead-associated FHA domain type-1 and -2, and adaptor protein 14-3-3 ligand motifs suggesting that the N-terminus of ataxin-1 may play a significant role in the assembly of the protein interactome.

6.1.1.4 Self association region (SAR)

At the cross-point between the N-terminal IUR and the AXH domain, a region of ca 100 aa (495-605) was identified in a yeast two hybrid system as responsible for protein self association (SAR) in cell (Burright et al., 1997). SAR shares 39 aa with the N-terminus of the AXH domain (567-689) which has been shown to be dimeric in solution (Chen et al., 2004; de Chiara et al., 2003). Therefore, this region seems to account for dimerization of the full-length ataxin-1. Interestingly, according to PONDR (VL3 predictor) and IUPRED, both based on the analysis of the local aminoacid composition, the full region ~440-700, which includes SAR and the AXH domain, is predicted as a potentially folded domain (**Figure 3**).

6.1.2 AXH domain

6.1.2.1 The structure of the ataxin-1 AXH domain

Soon after gene identification (Banfi et al., 1994), the analysis of the ataxin-1 sequence and the prediction of the secondary structure performed from multiple alignment of the protein from different species allowed the discovery of a new small putative independently folded domain (ca. ~130 aa) (with predicted predominantly beta structure) (de Chiara et al., 2003). The domain, successively named AXH (for Ataxin-1 Homology domain), did not show any detectable homology with any other known folding units (SMART accession number SM00536; http://smart.embl-heidelberg.de/) (Letunic et al., 2009; Schultz et al., 1998). A few years later, the homology between the ataxin-1 AXH and a region of an unrelated protein, the transcription factor HBP1, was detected (Mushegian et al., 1997). The two proteins share ~28% identity and ~54% similarity with the HBP1 AXH domain showing a ca. 10 aminoacids insertion loop between secondary structures respect to the ataxin-1 domain (de Chiara et al., 2003) (**Figure 4**).

The structure of ataxin-1 AXH, as solved by X-ray crystallography, consists of a non-canonical oligonucleotide- and oligosaccharide-binding (OB) fold (Chen et al., 2004; Murzin, 1993) (**Figure 5**). The AXH appears as a constitutive asymmetric dimer which crystallizes as an asymmetric dimer of dimers. Each monomer displays a common structure in the C-terminal part (residues 610-685), recognizable as the OB-fold, which superposes with an average root mean square deviation of 0.90 ± 0.06 Å on the backbone atoms. Conversely, approximately the first 30 N-terminal aminoacids show appreciable main chain differences between each of the two monomers in the dimer, with the same stretch of aminoacids adopting alternative secondary structures in the two cases. In this respect, the AXH domain represents an interesting example of a chameleon protein, which is a protein that adopts different folds under different environments. Interestingly, the observed structural differences in ataxin-1 are not induced by different experimental conditions or by the presence of ligands. Instead, they are present in the context of the same protein.

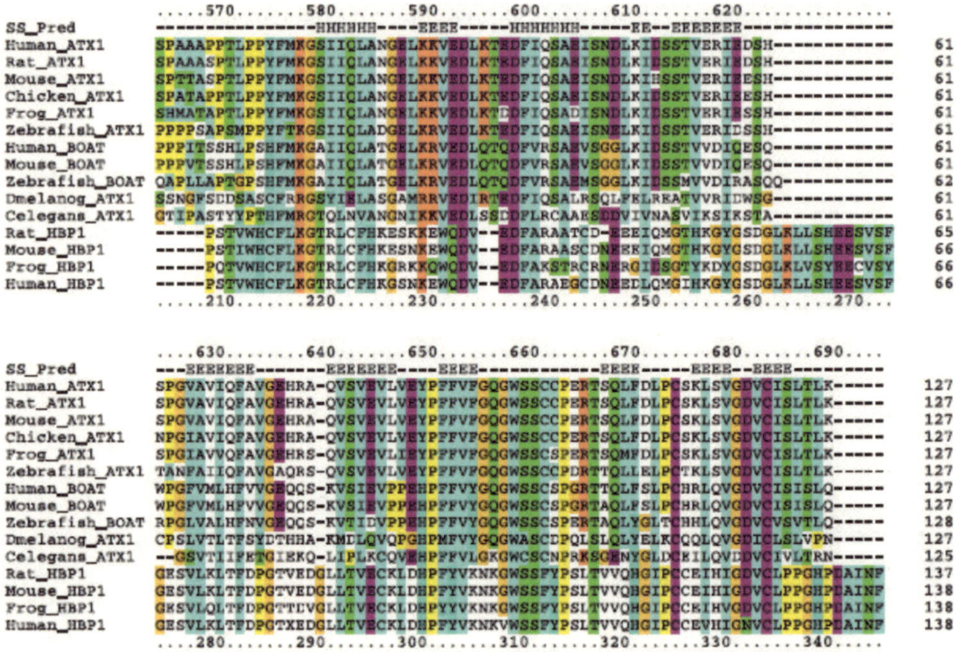

Fig. 4. Sequence alignment of the AXH from ataxin-1 (ATX1), ataxin-1 paralogue BOAT (Brother Of ATaxin-1) and HBP1 from different species. The alignment was prepared by ClustalX (version 2) (Larkin et al., 2007) and is based exclusively on sequence similarity. The secondary structure of the ataxin-1 AXH domain as predicted by Jpred 3 (Cole et al., 2008) is shown on the top for reference.

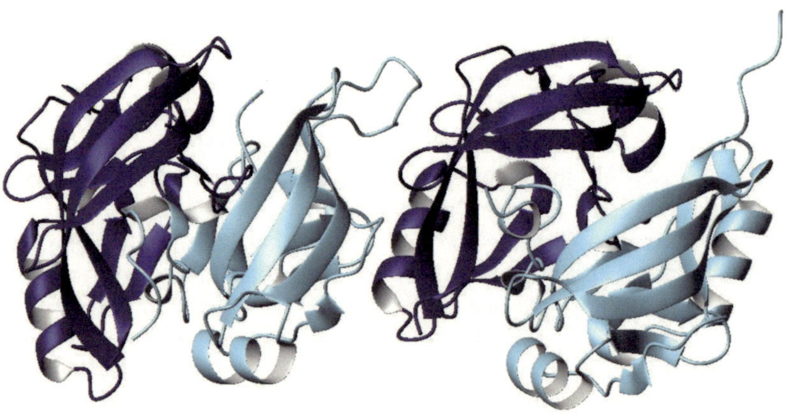

Fig. 5. X-ray structure of the AXH domain of ataxin-1 (PDB entry 1OA8) (Chen et al., 2004). The monomers forming the dimer of dimers observed in the structure are alternatively indicated in dark and light blue. Detailed analysis shows that they are not related by symmetry.

An even bigger surprise was revealed by structure determination of the AXH domain of HBP1. Whilst on the pure basis of sequence homology it would have been reasonable to assume that the structure of each of the two domains could be easily modelled from the other used as a template, an unexpected result came out when the solution structure of the HBP1 AXH domain was solved by NMR spectroscopy (de Chiara et al., 2005a) (**Figures 6 and 7**).

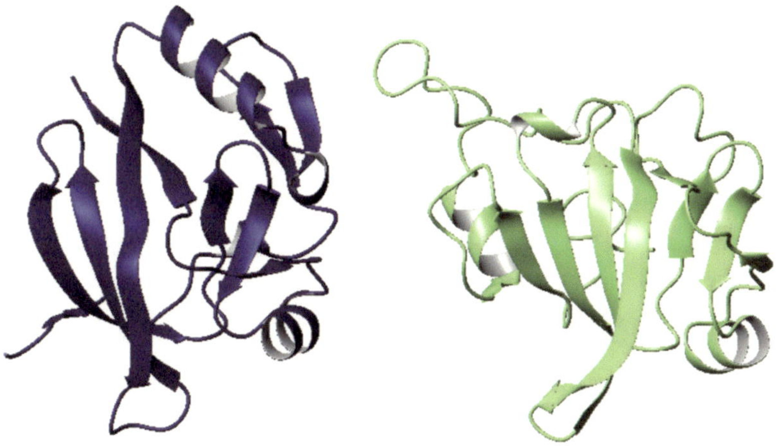

Fig. 6. Comparison between the structure of the AXH domains of ataxin-1 (monomer A) (left) (PDB entry 1OA8) (Chen et al., 2004) and HBP1 (right) (PDB entry 1V06) (de Chiara et al., 2005a). The N-terminal of the two monomers show the same elements of secondary structure arranged in a different topology.

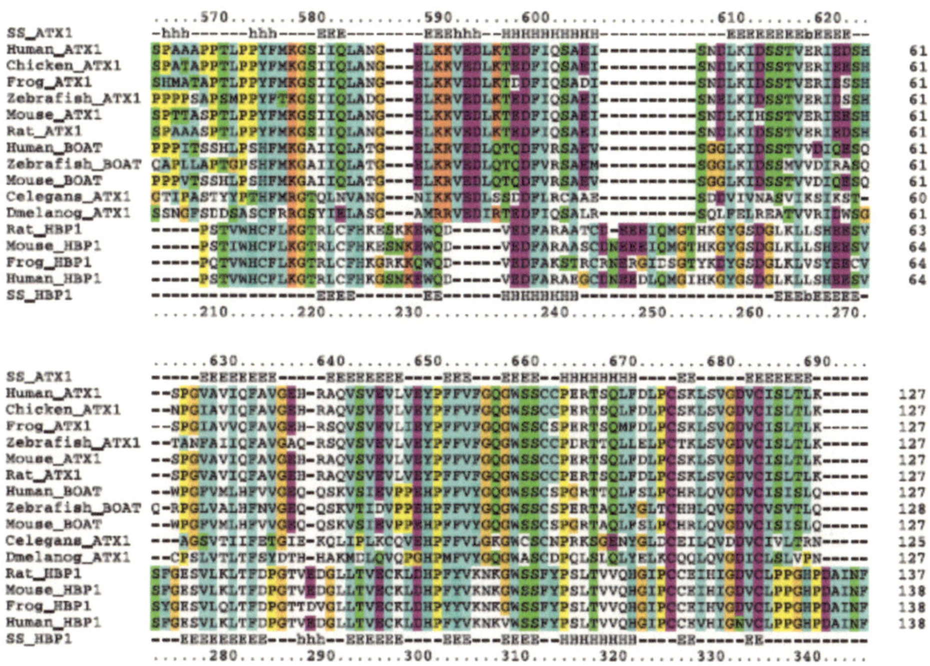

Fig. 7. Alignment based on structural comparison between the AXH from taxin-1 and HBP1 (Chen et al., 2004; de Chiara et al., 2005a). SS_ATX1 and SS_HBP1:experimental secondary structure of ataxin-1 and HBP1.

When comparing the structure of the HBP1 domain, which is monomeric (de Chiara et al., 2003; de Chiara et al., 2005a), to any of the ataxin-1 X-ray monomers, only the C-terminal part, representing the core of the OB-fold, is structurally superposable, while the N-terminus adopts a different topology, despite maintaining the same secondary structure elements along the sequence (**Figure 6**). Only a structure-based comparison allowed us to realign the sequences and to correctly position the HBP1 long-loop insertion, which was originally set between beta-3 and beta-4, between helix-1 and beta-3 (**Figure 7**). These findings support the possibility that the AXH motif is intrinsically able to adopt different topologies.

6.1.2.2 The function of the ataxin-1 AXH domain

Further studies on the role of the OB-fold of the AXH domain have shown that this region is designed to mediate interactions both with nucleic acids and proteins. The crystal structure of the ataxin-1 AXH allowed us to rationalize previous literature notions on the ability of the AXH to bind RNA homopolimers in vitro with the same nucleotide preference as full-length ataxin-1 (de Chiara et al., 2003; Yue et al., 2001). In addition to a direct binding to RNA through the AXH domain, the protein was found to co-localize with RNA also when the AXH domain was deleted, thus suggesting an involvement of other RNA binding proteins in the ataxin-1 interactome (de Chiara et al., 2005b). Recent findings on the ability of ataxin-1 to interact with splicing factors through a short motif localized C-terminally to the AXH domain opened the intriguing possibility that the protein may be involved in pre-mRNA processing at the level of the splicing machinery (de Chiara et al., 2009; Lim et al., 2008). However, no RNA targets have, as yet, been identified and more research is needed to address the question of whether the protein may play a role in RNA metabolism and/or nuclear RNA export as suggested also by co-localization with the mRNA export factor TAP/NXF1 (Irwin et al., 2005).
As for the ability of the AXH domain to mediate protein-protein interactions, several binding partners with transcriptional activity have been identified whose interaction with ataxin-1 is abolished when the AXH domain is deleted: the silencing mediator of retinoid and thyroid hormone receptors SMRT/SMRTER (Tsai et al., 2004), the repressor Capicua (Lam et al., 2006), the transcription factors Sensless/Gfi-1 (Tsuda et al., 2005) and Sp1 (Goold et al., 2007). A potential role for ataxin-1 in transcriptional regulation was suggested at a very early stage by the homology with HBP1 (Sampson et al., 2001; Tevosian et al., 1997). A general read-out assay for repression of transcription (de Chiara et al., 2005b) confirmed that the AXH domain represses transcription when tethered to DNA similarly to what was observed for full-length ataxin-1 (Tsai et al., 2004). However, cross-linking experiments showed that there is no direct binding between DNA and AXH domain indicating that the interaction is mediated by other co-transcriptional regulators (de Chiara et al., 2005b), as also confirmed experimentally later on (Bolger et al., 2007; Goold et al., 2007; Lam et al., 2006; Serra et al., 2006; Tsuda et al., 2005).

6.1.3 The C-terminal IUR

The region downstream to the AXH domain up to the C-terminal end of the protein (amino acids 690-816) represents an example of possible conflict between the results of different predictors. Disopred2, PONDR, IUPRED and DisEMBL predict the C-terminus as an almost completely disordered region **(Figures 3)**. According to GlobPlot (http://globplot.embl.de), which is based on the Russell/Linding scale (Linding et al., 2003b), the region 703-786 is predicted as a potential globular domain. There is also no agreement with the prediction from the SMART server which, instead, identifies the AXH as the only folded region in the protein (**Figures 2**). While still awaiting for a systematic experimental validation, we can already comment on these results in light of our findings.

6.1.3.1 Prediction of ELMs in the C-terminal IUR: A three-way molecular switch in ataxin-1 C-terminus

Consistent with the presence of disorder, several linear motifs were predicted in the protein C-terminus. Among these three overlapping linear motifs identified downstream to AXH were experimentally verified: a nuclear localization signal (NLS) (771-774) (Klement et al., 1998), a 14-3-3 binding motif (774-778: key conservation RxxSxP) (Chen et al., 2003) and a UHM ligand motif (ULM) (771-776) (de Chiara et al., 2009), present in proteins associated with splicing. These motifs represent a three-way molecular switch which plays an important role both for the function of the native protein and for pathogenesis. In addition to the expansion of the polyQ tract, nuclear localization is a strict requirement for the development of the pathology. Expanded ataxin-1 with mutated NLS fails to enter the nucleus and does not cause aggregation that is the typical phenotypic hallmark of the SCA1 pathology (Klement et al., 1998). Further to polyQ expansion and nuclear localization, phosphorylation of S776 has been identified as a condition necessary for development of SCA1 (Emamian et al., 2003). Phosphorylation of S776 has been confirmed to occur in vivo (Emamian et al., 2003; Jorgensen et al., 2009) and is required for recognition of ataxin-1 by the protein 14-3-3, a molecular adaptor which modulates, in a phosphorylation-dependent manner, the function of different proteins in their specific context (Chen et al., 2003). Mutation of S776 to an alanine in expanded ataxin-1, despite not affecting nuclear localization, prevented the development of the SCA1 phenotype (Emamian et al., 2003).

Recently, an UHM ligand motif (ULM) predicted by the ELM server in the C-terminus of the ataxin-1 sequence has been experimentally validated and characterized (de Chiara et al., 2009). The ULM motif was first identified in the splicing factors SF1 and SAP155 and shown to bind the UHM domain of U2AF65 (Corsini et al., 2007). Ataxin-1 ULM (771-776) strongly overlaps with the 14-3-3 ligand motif (774-778). However, whilst phosphorylation of S776 is crucial for recognition by 14-3-3, it only marginally affects the interaction with U2AF65, increasing the dissociation constant by only ~3 folds. Being the K_d between ataxin-1 S776-phosphorylated ULM and 14-3-3 (ζ isoform) two orders of magnitude smaller than U2AF65 (0.4 μM versus 36 μM) it was possible to conclude that, when S776 is phosphorylated, 14-3-3 is able to displace U2AF65. Under these conditions, the 14-3-3-bound expanded ataxin-1 is prone to aggregation. The S776A mutation, which hampers the interaction with 14-3-3, still allows the interaction with UHM domain of U2AF65 and potentially other splicing factors. These interactions, likely because of the extended dimension of the spliceosome complex, may play a protective role and prevent aggregation. Our findings allowed us to conclude that phosphorylation of S776 provides the switch that regulates binding of ataxin-1 to the protein 14-3-3 and components of the spliceosome, and suggests that pathology develops when aggregation competes with native interactions (**Figure 8**). This example also shows how the investigation of the native function of the polyQ proteins have provided valuable hints for understanding the molecular mechanisms of pathogenesis.

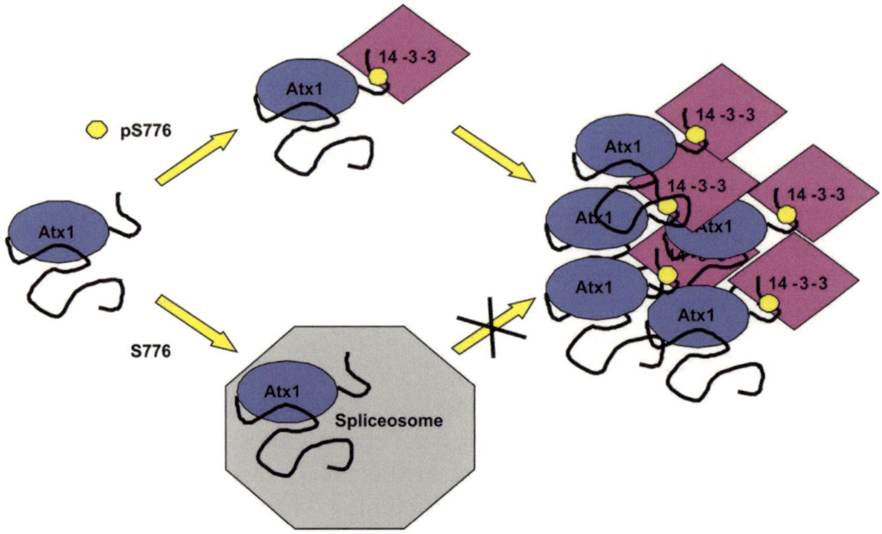

Fig. 8. A model of the role played by phosphorylation of S776 in the modulation of expanded ataxin-1 aggregation.

7. Conclusions

We have discussed here the concept of IUPs and IURs and shown how their detailed bioinformatics analysis can assist structural and functional studies using ataxin-1 as a paradigmatic example. Ataxin-1, like other members of the polyQ pathology family, is mostly composed of IUPs. Very little is still known about this protein despite its involvement in human neurodegeneration, yet this knowledge is essential for designing specific therapeutic interventions. Identification of both structured (the AXH domain) and unstructured linear functional motifs has played a key role in advancing our knowledge in the ataxin-1 function in the cell. More advanced information will undoubtedly come from experimental analysis of long stretches of the protein if the formidable challenges of their recombinant production in a pure and stable form can be circumvented. It is also evident from the example reported here how new approaches in the identification and study of IUPs might be highly helpful to advance the field.

8. References

Banfi, S., Servadio, A., Chung, M.Y., Kwiatkowski, T.J., Jr., McCall, A.E., Duvick, L.A., Shen, Y., Roth, E.J., Orr, H.T., & Zoghbi, H.Y. (1994). Identification and Characterization of the Gene Causing Type 1 Spinocerebellar Ataxia. *Nature Genetics*, Vol. 7, No. 4, (August 1994), pp. 513-520, ISSN 1061-4036

Bernado, P., Mylonas, E., Petoukhov, M.V., Blackledge, M., & Svergun, D.I. (2007). Structural Characterization of Flexible Proteins Using Small-Angle X-Ray Scattering. *Journal of the American Chemical Society*, Vol. 129, No. 17, (May 2007), pp. 5656-5664, ISSN 0002-7863

Boesch, C., Bundi, A., Oppliger, M., & Wuthrich, K. (1978). 1h Nuclear-Magnetic-Resonance Studies of the Molecular Conformation of Monomeric Glucagon in Aqueous

Solution. *European Journal of Biochemistry,* Vol. 91, No. 1, (November 1978), pp. 209-214, ISSN 0014-2956

Bolger, T.A., Zhao, X., Cohen, T.J., Tsai, C.C., &Yao, T.P. (2007). The Neurodegenerative Disease Protein Ataxin-1 Antagonizes the Neuronal Survival Function of Myocyte Enhancer Factor-2. *Journal of Biologic Chemistry,* Vol. 282, No. 40, (October 2007), pp. 29186-29192, ISSN 0021-9258

Bonvin, A.M., & Brunger, A.T. (1996). Do Noe Distances Contain Enough Information to Assess the Relative Populations of Multi-Conformer Structures? *Journal of Biomolecular NMR,* Vol. 7, No. 1, (January 1996), pp. 72-76, ISSN 0925-2738

Burright, E.N., Davidson, J.D., Duvick, L.A., Koshy, B., Zoghbi, H.Y., & Orr, H.T. (1997). Identification of a Self-Association Region within the Sca1 Gene Product, Ataxin-1. *Human Molecular Genetics,* Vol. 6, No. 4, (April 1997), pp. 513-518, ISSN 0964-6906

Chen, H.K., Fernandez-Funez, P., Acevedo, S.F., Lam, Y.C., Kaytor, M.D., Fernandez, M.H., Aitken, A., Skoulakis, E.M., Orr, H.T., Botas, J., & Zoghbi, H.Y. (2003). Interaction of Akt-Phosphorylated Ataxin-1 with 14-3-3 Mediates Neurodegeneration in Spinocerebellar Ataxia Type 1. *Cell,* Vol. 113, No. 4, (May 2003), pp. 457-468, ISSN 0092-8674

Chen, Y.W., Allen, M.D., Veprintsev, D.B., Lowe, J., & Bycroft, M. (2004). The Structure of the Axh Domain of Spinocerebellar Ataxin-1. *Journal of Biologic Chemistry,* Vol. 279, No. 5, (January 2004), pp. 3758-3765, ISSN 0021-9258

Clore, G.M., & Schwieters, C.D. (2004a). How Much Backbone Motion in Ubiquitin Is Required to Account for Dipolar Coupling Data Measured in Multiple Alignment Media as Assessed by Independent Cross-Validation? *Journal of the American Chemical Society,* Vol. 126, No. 9, (March 2004a), pp. 2923-2938, ISSN 0002-7863

Clore, G.M., & Schwieters, C.D. (2004b). Amplitudes of Protein Backbone Dynamics and Correlated Motions in a Small Alpha/Beta Protein: Correspondence of Dipolar Coupling and Heteronuclear Relaxation Measurements. *Biochemistry,* Vol. 43, No. 33, (Aug 24 2004b), pp. 10678-10691, ISSN 0006-2960

Cole, C., Barber, J.D., & Barton, G.J. (2008). The Jpred 3 Secondary Structure Prediction Server. *Nucleic Acids Research,* Vol. 36, Web Server issue, (July 2008), pp. W197-201, ISSN 1362-4962

Corsini, L., Bonnal, S., Basquin, J., Hothorn, M., Scheffzek, K., Valcarcel, J., & Sattler, M. (2007). U2af-Homology Motif Interactions Are Required for Alternative Splicing Regulation by Spf45. *Nature Structure and Molecular Biology,* Vol. 14, No. 7, (July 2007), pp. 620-629, ISSN 1545-9993

Csizmok, V., Dosztanyi, Z., Simon, I., & Tompa, P. (2007). Towards Proteomic Approaches for the Identification of Structural Disorder. *Current Protein & Peptide Science,* Vol. 8, No. 2, (April 2007), pp. 173-179, ISSN 1389-2037

de Chiara, C., Giannini, C., Adinolfi, S., de Boer, J., Guida, S., Ramos, A., Jodice, C., Kioussis, D., & Pastore, A. (2003). The Axh Module: An Independently Folded Domain Common to Ataxin-1 and Hbp1. *FEBS Letters,* Vol. 551, No. 1-3, (September 2003), pp. 107-112, ISSN 0014-5793

de Chiara, C., Menon, R.P., Adinolfi, S., de Boer, J., Ktistaki, E., Kelly, G., Calder, L., Kioussis, D., & Pastore, A. (2005a). The Axh Domain Adopts Alternative Folds the Solution Structure of Hbp1 Axh. *Structure,* Vol. 13, No. 5, (May 2005a), pp. 743-753, ISSN 0969-2126 de Chiara, C., Menon, R.P., Dal Piaz, F., Calder, L., &Pastore, A. (2005b). Polyglutamine Is Not All: The Functional Role of the Axh Domain in the Ataxin-1 Protein. *Journal of Molecular Biology,* Vol. 354, No. 4, (December 2005b), pp. 883-893, ISSN 0022-2836

de Chiara, C., Menon, R.P., Strom, M., Gibson, T.J., & Pastore, A. (2009). Phosphorylation of S776 and 14-3-3 Binding Modulate Ataxin-1 Interaction with Splicing Factors. *PLoS One,* Vol. 4, No. 12, (December 2009), pp. e8372, ISSN 1932-6203

Dephoure, N., Zhou, C., Villen, J., Beausoleil, S.A., Bakalarski, C.E., Elledge, S.J., & Gygi, S.P. (2008). A Quantitative Atlas of Mitotic Phosphorylation. *Proceedings of the National Academy of Sciences U S A,* Vol. 105, No. 31, (August 2008), pp. 10762-10767, ISSN 1091-6490

Diella, F., Haslam, N., Chica, C., Budd, A., Michael, S., Brown, N.P., Trave, G., & Gibson, T.J. (2008). Understanding Eukaryotic Linear Motifs and Their Role in Cell Signaling and Regulation. *Frontiers in Bioscience,* Vol. 13, No. (May 2008), pp. 6580-6603, ISSN 1093-4715

Dosztanyi, Z., Csizmok, V., Tompa, P., & Simon, I. (2005a). The Pairwise Energy Content Estimated from Amino Acid Composition Discriminates between Folded and Intrinsically Unstructured Proteins. *Journal of Molecular Biology,* Vol. 347, No. 4, (April 2005a), pp. 827-839, ISSN 0022-2836

Dosztanyi, Z., Csizmok, V., Tompa, P., & Simon, I. (2005b). Iupred: Web Server for the Prediction of Intrinsically Unstructured Regions of Proteins Based on Estimated Energy Content. *Bioinformatics,* Vol. 21, No. 16, (August 2005b), pp. 3433-3434, ISSN 1367-4803

Dunker, A.K., Lawson, J.D., Brown, C.J., Williams, R.M., Romero, P., Oh, J.S., Oldfield, C.J., Campen, A.M., Ratliff, C.M., Hipps, K.W., Ausio, J., Nissen, M.S., Reeves, R., Kang, C., Kissinger, C.R., Bailey, R.W., Griswold, M.D., Chiu, W., Garner, E.C., & Obradovic, Z. (2001). Intrinsically Disordered Protein. *Journal of Molecular Graphics and Modelling,* Vol. 19, No. 1, pp. 26-59, ISSN 1093-3263

Dunker, A.K., Brown, C.J., Lawson, J.D., Iakoucheva, L.M., & Obradovic, Z. (2002). Intrinsic Disorder and Protein Function. *Biochemistry,* Vol. 41, No. 21, (May 2002), pp. 6573-6582, ISSN 0006-2960

Emamian, E.S., Kaytor, M.D., Duvick, L.A., Zu, T., Tousey, S.K., Zoghbi, H.Y., Clark, H.B., & Orr, H.T. (2003). Serine 776 of Ataxin-1 Is Critical for Polyglutamine-Induced Disease in Sca1 Transgenic Mice. *Neuron,* Vol. 38, No. 3, (May 2003), pp. 375-387, ISSN 0896-6273

Evans, P.R., & Owen, D.J. (2002). Endocytosis and Vesicle Trafficking. *Current Opinion in Structural Biology,* Vol. 12, No. 6, (December 2002), pp. 814-821, ISSN 0959-440X

Fennen, J., Torda, A.E., & van Gunsteren, W.F. (1995). Structure Refinement with Molecular Dynamics and a Boltzmann-Weighted Ensemble. *Journal of Biomolecular NMR,* Vol. 6, No. 2, (September 1995), pp. 163-170, ISSN 0925-2738

Ferron, F., Longhi, S., Canard, B., & Karlin, D. (2006). A Practical Overview of Protein Disorder Prediction Methods. *Proteins,* Vol. 65, No. 1, (October 2006), pp. 1-14, ISSN 1097-0134

Finn, R.D., Mistry, J., Tate, J., Coggill, P., Heger, A., Pollington, J.E., Gavin, O.L., Gunasekaran, P., Ceric, G., Forslund, K., Holm, L., Sonnhammer, E.L., Eddy, S.R., & Bateman, A. (2010). The Pfam Protein Families Database. *Nucleic Acids Research,* Vol. 38, Database issue, (January 2010), pp. D211-222, ISSN 1362-4962

Genis, D., Matilla, T., Volpini, V., Rosell, J., Davalos, A., Ferrer, I., Molins, A., & Estivill, X. (1995). Clinical, Neuropathologic, and Genetic Studies of a Large Spinocerebellar Ataxia Type 1 (Sca1) Kindred: (Cag)N Expansion and Early Premonitory Signs and Symptoms. *Neurology,* Vol. 45, No. 1, (January 1995), pp. 24-30, ISSN 0028-3878

Gibson, T.J. (2009). Cell Regulation: Determined to Signal Discrete Cooperation. *Trends in Biochemical Sciences,* Vol. 34, No. 10, (October 2009), pp. 471-482, ISSN 0968-0004

Goold, R., Hubank, M., Hunt, A., Holton, J., Menon, R.P., Revesz, T., Pandolfo, M., & Matilla-Duenas, A. (2007). Down-Regulation of the Dopamine Receptor D2 in Mice Lacking Ataxin 1. *Human Molecular Genetics,* Vol. 16, No. 17, (September 2007), pp. 2122-2134, ISSN 0964-6906

Greaser, M.L., Wang, S.M., Berri, M., Mozdziak, P., & Kumazawa, Y. (2000). Sequence and Mechanical Implications of Titin's Pevk Region. *Advances in Experimental Medicine and Biology,* Vol. 481, pp. 53-63; discussion 64-56, 107-110, ISSN 0065-2598

Iakoucheva, L.M., Brown, C.J., Lawson, J.D., Obradovic, Z., & Dunker, A.K. (2002). Intrinsic Disorder in Cell-Signaling and Cancer-Associated Proteins. *Journal of Molecular Biology,* Vol. 323, No. 3, (October 2002), pp. 573-584, ISSN 0022-2836

Irwin, S., Vandelft, M., Pinchev, D., Howell, J.L., Graczyk, J., Orr, H.T., & Truant, R. (2005). Rna Association and Nucleocytoplasmic Shuttling by Ataxin-1. *Journal of Cell Science,* Vol. 118, No. Pt 1, (January 2005), pp. 233-242, ISSN 0021-9533

Jayaraman, M., Kodali, R., & Wetzel, R. (2009). The Impact of Ataxin-1-Like Histidine Insertions on Polyglutamine Aggregation. *Protein Engineering Design Selection,* Vol. 22, No. 8, (August 2009), pp. 469-478, ISSN 1741-0134

Jorgensen, N.D., Andresen, J.M., Lagalwar, S., Armstrong, B., Stevens, S., Byam, C.E., Duvick, L.A., Lai, S., Jafar-Nejad, P., Zoghbi, H.Y., Clark, H.B., & Orr, H.T. (2009). Phosphorylation of Atxn1 at Ser776 in the Cerebellum. *Journal of Neurochemistry,* Vol. 110, No. 2, (July 2009), pp. 675-686, ISSN 1471-4159

Klement, I.A., Skinner, P.J., Kaytor, M.D., Yi, H., Hersch, S.M., Clark, H.B., Zoghbi, H.Y., & Orr, H.T. (1998). Ataxin-1 Nuclear Localization and Aggregation: Role in Polyglutamine-Induced Disease in Sca1 Transgenic Mice. *Cell,* Vol. 95, No. 1, (October 1998), pp. 41-53, ISSN 0092-8674

Konagurthu, A.S., &Lesk, A.M. Cataloging Topologies of Protein Folding Patterns. *J Mol Recognit,* Vol. 23, No. 2, (March-April 2010) pp. 253-257, ISSN 1099-1352

Kuszewski, J., Gronenborn, A.M., & Clore, G.M. (1996). Improving the Quality of Nmr and Crystallographic Protein Structures by Means of a Conformational Database Potential Derived from Structure Databases. *Protein Science,* Vol. 5, No. 6, (June 1996), pp. 1067-1080, ISSN 0961-8368

Labeit, S., &Kolmerer, B. (1995). Titins: Giant Proteins in Charge of Muscle Ultrastructure and Elasticity. *Science,* Vol. 270, No. 5234, (October 1995), pp. 293-296, ISSN 0036-8075

Lam, Y.C., Bowman, A.B., Jafar-Nejad, P., Lim, J., Richman, R., Fryer, J.D., Hyun, E.D., Duvick, L.A., Orr, H.T., Botas, J., & Zoghbi, H.Y. (2006). Ataxin-1 Interacts with the Repressor Capicua in Its Native Complex to Cause Sca1 Neuropathology. *Cell,* Vol. 127, No. 7, (December 2006), pp. 1335-1347, ISSN 0092-8674

Larkin, M.A., Blackshields, G., Brown, N.P., Chenna, R., McGettigan, P.A., McWilliam, H., Valentin, F., Wallace, I.M., Wilm, A., Lopez, R., Thompson, J.D., Gibson, T.J., &

Higgins, D.G. (2007). Clustal W and Clustal X Version 2.0. *Bioinformatics,* Vol. 23, No. 21, (November 2007), pp. 2947-2948, ISSN 1367-4811

Letunic, I., Doerks, T., & Bork, P. (2009). Smart 6: Recent Updates and New Developments. *Nucleic Acids Research,* Vol. 37, Database issue, (January 2009), pp. D229-232, ISSN 1362-4962

Lim, J., Crespo-Barreto, J., Jafar-Nejad, P., Bowman, A.B., Richman, R., Hill, D.E., Orr, H.T., & Zoghbi, H.Y. (2008). Opposing Effects of Polyglutamine Expansion on Native Protein Complexes Contribute to Sca1. *Nature,* Vol. 452, No. 7188, (April 2008), pp. 713-718, ISSN 1476-4687

Linding, R., Jensen, L.J., Diella, F., Bork, P., Gibson, T.J., & Russell, R.B. (2003a). Protein Disorder Prediction: Implications for Structural Proteomics. *Structure,* Vol. 11, No. 11, (November 2003a), pp. 1453-1459, ISSN 0969-2126

Linding, R., Russell, R.B., Neduva, V., & Gibson, T.J. (2003b). Globplot: Exploring Protein Sequences for Globularity and Disorder. *Nucleic Acids Research,* Vol. 31, No. 13, (July 2003b), pp. 3701-3708, ISSN 1362-4962

Masino, L., Kelly, G., Leonard, K., Trottier, Y., & Pastore, A. (2002). Solution Structure of Polyglutamine Tracts in Gst-Polyglutamine Fusion Proteins. *FEBS Letters,* Vol. 513, No. 2-3, (February 2002), pp. 267-272, ISSN 0014-5793

Mittag, T., & Forman-Kay, J.D. (2007). Atomic-Level Characterization of Disordered Protein Ensembles. *Current Opinion in Structural Biology,* Vol. 17, No. 1, (February 2007), pp. 3-14, ISSN 0959-440X

Murzin, A.G. (1993). Ob(Oligonucleotide/Oligosaccharide Binding)-Fold: Common Structural and Functional Solution for Non-Homologous Sequences. *The EMBO Journal,* Vol. 12, No. 3, (March 1993), pp. 861-867, ISSN 0261-4189

Mushegian, A.R., Bassett, D.E., Jr., Boguski, M.S., Bork, P., & Koonin, E.V. (1997). Positionally Cloned Human Disease Genes: Patterns of Evolutionary Conservation and Functional Motifs. *Proceedings of the National Academy of Sciences U.S.A.,* Vol. 94, No. 11, (May 1997), pp. 5831-5836, ISSN 0027-8424

Okazawa, H., Rich, T., Chang, A., Lin, X., Waragai, M., Kajikawa, M., Enokido, Y., Komuro, A., Kato, S., Shibata, M., Hatanaka, H., Mouradian, M.M., Sudol, M., & Kanazawa, I. (2002). Interaction between Mutant Ataxin-1 and Pqbp-1 Affects Transcription and Cell Death. *Neuron,* Vol. 34, No. 5, (May 2002), pp. 701-713, ISSN 0896-6273

Orr, H.T., Chung, M.Y., Banfi, S., Kwiatkowski, T.J., Jr., Servadio, A., Beaudet, A.L., McCall, A.E., Duvick, L.A., Ranum, L.P., & Zoghbi, H.Y. (1993). Expansion of an Unstable Trinucleotide Cag Repeat in Spinocerebellar Ataxia Type 1. *Nature Genetics,* Vol. 4, No. 3, (July 1993), pp. 221-226, ISSN 1061-4036

Orr, H.T., & Zoghbi, H.Y. (2007). Trinucleotide Repeat Disorders. *Annual Review in Neuroscience,* Vol. 30, pp. 575-621, ISSN 0147-006X

Puntervoll, P., Linding, R., Gemund, C., Chabanis-Davidson, S., Mattingsdal, M., Cameron, S., Martin, D.M., Ausiello, G., Brannetti, B., Costantini, A., Ferre, F., Maselli, V., Via, A., Cesareni, G., Diella, F., Superti-Furga, G., Wyrwicz, L., Ramu, C., McGuigan, C., Gudavalli, R., Letunic, I., Bork, P., Rychlewski, L., Kuster, B., Helmer-Citterich, M., Hunter, W.N., Aasland, R., & Gibson, T.J. (2003). Elm Server: A New Resource for Investigating Short Functional Sites in Modular Eukaryotic Proteins. *Nucleic Acids Research,* Vol. 31, No. 13, (July 2003), pp. 3625-3630, ISSN 1362-4962

Radivojac, P., Obradovic, Z., Brown, C.J., & Dunker, A.K. (2003). Prediction of Boundaries between Intrinsically Ordered and Disordered Protein Regions. *Pacific Symposium on Biocomputing*, Vol. 8, pp. 216-227, ISSN 1793-5091

Radivojac, P., Iakoucheva, L.M., Oldfield, C.J., Obradovic, Z., Uversky, V.N., & Dunker, A.K. (2007). Intrinsic Disorder and Functional Proteomics. *Biophysical Journal*, Vol. 92, No. 5, (March 2007), pp. 1439-1456, ISSN 1542-0086

Romero, P., Pbradovic, Z., Kissinger, C.R., Villafranca, J.E., & Dunker, A.K. Identifying Disordered Regions in Proteins from Amino Acid Sequences; 1997; Houston, TX. pp. 90-95.

Sampson, E.M., Haque, Z.K., Ku, M.C., Tevosian, S.G., Albanese, C., Pestell, R.G., Paulson, K.E., & Yee, A.S. (2001). Negative Regulation of the Wnt-Beta-Catenin Pathway by the Transcriptional Repressor Hbp1. *The Embo Journal*, Vol. 20, No. 16, (August 2001), pp. 4500-4511, ISSN 0261-4189

Schaeffer, R.D., &Daggett, V. Protein Folds and Protein Folding. *Protein Engineering Design and Selection*, Vol. 24, No. 1-2, (January 2010), pp. 11-19, ISSN 1741-0134

Schultz, J., Milpetz, F., Bork, P., & Ponting, C.P. (1998). Smart, a Simple Modular Architecture Research Tool: Identification of Signaling Domains. *Proceedings of the National Academy of Sciences U S A*, Vol. 95, No. 11, (May 1998), pp. 5857-5864, ISSN 0027-8424

Serra, H.G., Duvick, L., Zu, T., Carlson, K., Stevens, S., Jorgensen, N., Lysholm, A., Burright, E., Zoghbi, H.Y., Clark, H.B., Andresen, J.M., &Orr, H.T. (2006). Roralpha-Mediated Purkinje Cell Development Determines Disease Severity in Adult Sca1 Mice. *Cell*, Vol. 127, No. 4, (November 2006), pp. 697-708, ISSN 0092-8674

Shoemaker, B.A., Portman, J.J., & Wolynes, P.G. (2000). Speeding Molecular Recognition by Using the Folding Funnel: The Fly-Casting Mechanism. *Proceedings of the National Academy of Sciences U S A*, Vol. 97, No. 16, (August 2000), pp. 8868-8873, ISSN 0027-8424

Sjekloća, L., Pauwels, K., & Pastore, A. (2011) *FEBS Journal*, Accepted Article; doi: 10.1111/j.1742-4658.2011.08108.x

Tevosian, S.G., Shih, H.H., Mendelson, K.G., Sheppard, K.A., Paulson, K.E., & Yee, A.S. (1997). Hbp1: A Hmg Box Transcriptional Repressor That Is Targeted by the Retinoblastoma Family. *Genes & Development*, Vol. 11, No. 3, (February 1997), pp. 383-396, ISSN 0890-9369

Tompa, P., & Csermely, P. (2004). The Role of Structural Disorder in the Function of Rna and Protein Chaperones. *Faseb Journal*, Vol. 18, No. 11, (August 2004), pp. 1169-1175, ISSN 1530-6860

Tompa, P. (2005). The Interplay between Structure and Function in Intrinsically Unstructured Proteins. *FEBS Letters*, Vol. 579, No. 15, (June 2005), pp. 3346-3354, ISSN 0014-5793

Tompa, P., & Fuxreiter, M. (2008). Fuzzy Complexes: Polymorphism and Structural Disorder in Protein-Protein Interactions. *Trends in Biochemical Sciences*, Vol. 33, No. 1, (January 2008), pp. 2-8, ISSN 0968-0004

Tsai, C.C., Kao, H.Y., Mitzutani, A., Banayo, E., Rajan, H., McKeown, M., & Evans, R.M. (2004). Ataxin 1, a Sca1 Neurodegenerative Disorder Protein, Is Functionally Linked to the Silencing Mediator of Retinoid and Thyroid Hormone Receptors.

Proceedings of the National Academy of Sciences U.S.A., Vol. 101, No. 12, (March 2004), pp. 4047-4052, ISSN 0027-8424

Tsuda, H., Jafar-Nejad, H., Patel, A.J., Sun, Y., Chen, H.K., Rose, M.F., Venken, K.J., Botas, J., Orr, H.T., Bellen, H.J., & Zoghbi, H.Y. (2005). The Axh Domain of Ataxin-1 Mediates Neurodegeneration through Its Interaction with Gfi-1/Senseless Proteins. *Cell,* Vol. 122, No. 4, (August 2005), pp. 633-644, ISSN 0092-8674

Uversky, V.N. (2002). What Does It Mean to Be Natively Unfolded? *European Journal of Biochemistry,* Vol. 269, No. 1, (January 2002), pp. 2-12, ISSN 0014-2956

Vendruscolo, M., & Paci, E. (2003). Protein Folding: Bringing Theory and Experiment Closer Together. *Current Opinion in Structural Biology,* Vol. 13, No. 1, (February 2003), pp. 82-87, ISSN 0959-440X

Vendruscolo, M., & Dobson, C.M. (2005). Towards Complete Descriptions of the Free-Energy Landscapes of Proteins. *Philos Transact A Math Phys Eng Sci,* Vol. 363, No. 1827, (February 2005), pp. 433-450; discussion 450-432, ISSN 1364-503X

Vierra-Green, C.A., Orr, H.T., Zoghbi, H.Y., & Ferrington, D.A. (2005). Identification of a Novel Phosphorylation Site in Ataxin-1. *Biochimica et Biophysica Acta,* Vol. 1744, No. 1, (May 2005), pp. 11-18, ISSN 0006-3002

Ward, J.J., McGuffin, L.J., Bryson, K., Buxton, B.F., & Jones, D.T. (2004a). The Disopred Server for the Prediction of Protein Disorder. *Bioinformatics,* Vol. 20, No. 13, (September 2004a), pp. 2138-2139, ISSN 1367-4803

Ward, J.J., Sodhi, J.S., McGuffin, L.J., Buxton, B.F., & Jones, D.T. (2004b). Prediction and Functional Analysis of Native Disorder in Proteins from the Three Kingdoms of Life. *Journal of Molecular Biology,* Vol. 337, No. 3, (March 2004b), pp. 635-645, ISSN 0022-2836

Wright, P.E., & Dyson, H.J. (1999). Intrinsically Unstructured Proteins: Re-Assessing the Protein Structure-Function Paradigm. *Journal of Molecular Biology,* Vol. 293, No. 2, (October 1999), pp. 321-331, ISSN 0022-2836

Wright, P.E., & Dyson, H.J. (2009). Linking Folding and Binding. *Current Opinion in Structural Biology,* Vol. 19, No. 1, (February 2009), pp. 31-38, ISSN 1879-033X

Yue, S., Serra, H.G., Zoghbi, H.Y., & Orr, H.T. (2001). The Spinocerebellar Ataxia Type 1 Protein, Ataxin-1, Has Rna-Binding Activity That Is Inversely Affected by the Length of Its Polyglutamine Tract. *Human Molecular Genetics,* Vol. 10, No. 1, (January 2001), pp. 25-30, ISSN 0964-6906

Zoghbi, H.Y., & Orr, H.T. (2008). Pathogenic Mechanisms of a Polyglutamine Mediated Neurodegenerative Disease: Sca1. *Journal of Biologic Chemistry,* Vol. 284, No. 12, (October 2008), pp. 7425-7429, ISSN 0021-9258

6

Relaxed Linear Separability (RLS) Approach to Feature (Gene) Subset Selection

Leon Bobrowski[1,2] and Tomasz Łukaszuk[1]
[1]*Faculty of Computer Science, Białystok University of Technology,*
[2]*Institute of Biocybernetics and Biomedical Engineering, PAS, Warsaw*
Poland

1. Introduction

Feature selection is one of active research area in pattern recognition or data mining methods (Duda et al., 2001). The importance of feature selection methods becomes apparent in the context of rapidly growing amount of data collected in contemporary databases (Liu & Motoda, 2008).

Feature subset selection procedures are aimed at neglecting as large as possible number of such features (measurements) which are irrelevant or redundant for a given problem. The feature subset resulting from feature selection procedure should allow to build a model on the base of available learning data sets that generalizes better to new (unseen) data. For the purpose of designing classification or prediction models, the feature subset selection procedures are expected to produce higher classification or prediction accuracy.

Feature selection problem is particularly important and challenging in the case when the number of objects represented in a given database is low in comparison to the number of features which have been used to characterise these objects. Such situation appears typically in exploration of genomic data sets where the number of features can be thousands of times greater than the number of objects.

Here we are considering the *relaxed linear separability* (*RLS*) method of feature subset selection (Bobrowski & Łukaszuk, 2009). Such approach to feature selection problem refers to the concept of linear separability of the learning sets (Bobrowski, 2008). The term "relaxation" means here deterioration of the linear separability due to the gradual neglect of selected features. The considered approach to feature selection is based on repetitive minimization of the convex and piecewise-linear (*CPL*) criterion functions. These *CPL* criterion functions, which have origins in the theory of neural networks, include the cost of various features (Bobrowski, 2005). Increasing the cost of individual features makes these features falling out of the feature subspace. Quality the reduced feature subspaces is assessed by the accuracy of the CPL optimal classifiers built in this subspace.

The article contains a new theoretical and experimental results related to the RLS method of feature subset selection. The experimental results have been achieved through the analysis, inter alia, two sets of genetic data.

2. Linear separability of two learning sets

Suppose that m objects O_j described in the database are represented by feature vectors $\mathbf{x}_j[n] = [x_{j1},...,x_{jn}]^T$ ($j = 1,\ldots,m$). The feature vectors $\mathbf{x}_j[n]$ can be treated as points in the n-dimensional feature space $F[n]$ ($\mathbf{x}_j[n] \in F[n]$). The component x_{ji} of the vector $\mathbf{x}_j[n]$ is the numerical value of the i-th feature x_i of the object O_j. For example, in the case of clinical database, components x_{ji} can be the numerical results of the i-th diagnostic examinations of a given patient O_j.

Consider two learning sets G^+ and G^- built from n-dimensional feature vectors $\mathbf{x}_j[n]$. The *positive set* G^+ contains m^+ feature vectors $\mathbf{x}_j[n]$ and the *negative set* G^- contains m^- vectors $\mathbf{x}_j[n]$:

$$G^+ = \{\mathbf{x}_j[n]: j \in J^+\} \quad \textit{\textbf{and}} \quad G^- = \{\mathbf{x}_j[n]: j \in J^-\} \tag{1}$$

where J^+ and J^- are disjoined sets ($J^+ \cap J^- = \emptyset$) of indices j.

The positive set G^+ usually contains vectors $\mathbf{x}_j[n]$ of only one category. For example, the set G^+ may contain feature vectors $\mathbf{x}_j[n]$ representing patients with cancer and set G^- may represent patients without cancer.

Definition 1: The sets G^+ and G^- (1) are linearly separable, if and only if there exists such a weight vector $\mathbf{w}[n] = [w_1,...,w_n]^T$ ($\mathbf{w}[n] \in R^n$) and threshold θ ($\theta \in R$), that all the below inequalities are fulfilled:

$$\begin{aligned} &(\exists\, \mathbf{w}[n], \theta)\ (\forall \mathbf{x}_j[n] \in G^+)\ \mathbf{w}[n]^T\mathbf{x}_j[n] > \theta \\ &\qquad \textit{\textbf{and}}\ (\forall \mathbf{x}_j[n] \in G^-)\ \mathbf{w}[n]^T\mathbf{x}_j[n] < \theta \end{aligned} \tag{2}$$

The parameters $\mathbf{w}[n]$ and θ define the separating hyperplane $H(\mathbf{w}[n],\theta)$ in the feature space $F[n]$ ($\mathbf{x}[n] \in F[n]$):

$$H(\mathbf{w}[n],\theta) = \{\mathbf{x}[n]: \mathbf{w}[n]^T\mathbf{x}[n] = \theta\} \tag{3}$$

If the relations (2) are fulfilled, then all the elements $\mathbf{x}_j[n]$ of the set G^+ are located on the positive side of the hyperplane $H(\mathbf{w}[n],\theta)$ (3) and all the elements of the set G^- are located on the negative side of this hyperplane.

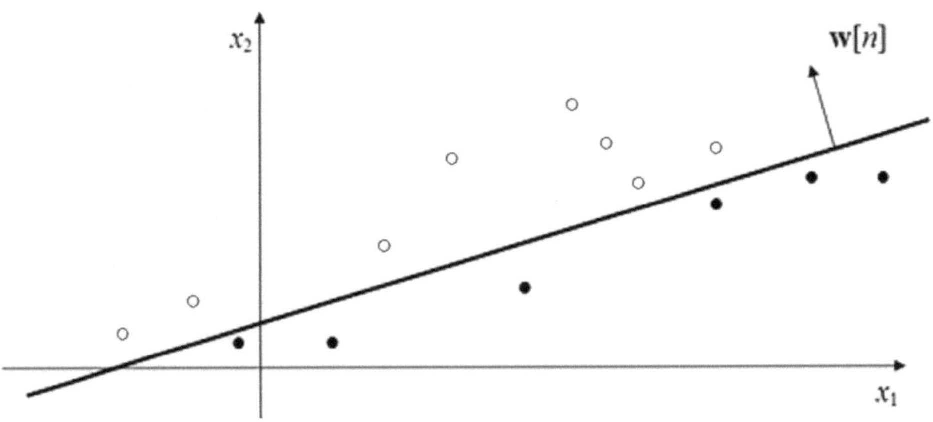

Fig. 1. An example of linearly separable sets G^+ (denoted by ○) and G^- (denoted by •) in the two-dimensional feature space F[2], where $m^+ = 8$ and $m^- = 6$

Lemma 1: Such sets G^+ and G^- (1) which are linearly separable (2) in the feature space $F[n]$, are also linearly separable in any greater feature space $F'[n']$, where $F[n] \subset F'[n']$.
The proof of the *Lemma* 1 is self-evident. The *Lemma* 1 shows, inter alia, that for any constant c the sets $G^+ = \{\mathbf{x}_j[n]: x_{ji'} > c\}$ and $G^- = \{\mathbf{x}_j[n]: x_{ji'} < c\}$ are linearly separable in each feature space $F[n]$.
Lemma 2: The sets G^+ and G^- (1) constructed of linearly independent feature vectors $\mathbf{x}_j[n]$ are always linearly separable (2) in the feature space $F[n]$.
The *Lemma* 2 can be proved by using arguments related to the construction of bases in the feature space $F[n]$ (Bobrowski, 2005). A base in the feature space $F[n]$ can be created by any n feature vectors $\mathbf{x}_j[n]$ which are linearly independent. Such n vectors $\mathbf{x}_j[n]$ can be separated by the hyperplane $H(\mathbf{w}[n],\theta)$ (3) for any subsets G^+ and G^- (1).
It can be seen that the linear separability (2) can be formulated equivalently to (2) as (Bobrowski, 2005):

$$\begin{aligned}(\exists\, \mathbf{v}[n+1])\ (\forall\, \mathbf{y}_j[n+1] \in G^+)\quad \mathbf{v}[n+1]^T\mathbf{y}_j[n+1] \geq 1 \\ \textit{and}\ (\forall\, \mathbf{y}_j[n+1] \in G^-)\quad \mathbf{v}[n+1]^T\mathbf{y}_j[n+1] \leq -1\end{aligned} \tag{4}$$

where $\mathbf{y}_j[n+1]$ are the *augmented feature vectors*, and $\mathbf{v}[n+1]$ is the *augmented weight vector* (Duda et al., 2001):

$$(\forall j \in \{1,\ldots,m\})\ \mathbf{y}_j[n+1] = [\mathbf{x}_j[n]^T, 1]^T \ \textit{and}\ \mathbf{v}[n+1] = [\mathbf{w}[n]^T, -\theta]^T \tag{5}$$

The inequalities (4) are used in the definition of the convex and piecewise-linear (*CPL*) penalty functions $\varphi_j^+(\mathbf{v}[n+1])$ and $\varphi_j^-(\mathbf{v}[n+1])$.

3. Convex and piecewise linear (*CPL*) criterion functions

Let us define the convex and piecewise-linear penalty functions $\varphi_j^+(\mathbf{v}[n+1])$ and $\varphi_j^-(\mathbf{v}[n+1])$ using the augmented feature vectors $\mathbf{y}_j[n+1]$ (5), and the weight vector $\mathbf{v}[n+1]$ (Bobrowski, 2005):

$$(\forall \mathbf{y}_j[n+1] \in G^+)\ \varphi_j^+(\mathbf{v}[n+1]) = \begin{matrix} 1 - \mathbf{v}[n+1]^T\mathbf{y}_j[n+1] & \textit{if} & \mathbf{v}[n+1]^T\mathbf{y}_j[n+1] < 1 \\ 0 & \textit{if} & \mathbf{v}[n+1]^T\mathbf{y}_j[n+1] \geq 1 \end{matrix} \tag{6}$$

and

$$(\forall \mathbf{y}_j[n+1] \in G^-)\ \varphi_j^-(\mathbf{v}[n+1]) = \begin{matrix} 1 + \mathbf{v}[n+1]^T\mathbf{y}_j[n+1] & \textit{if} & \mathbf{v}[n+1]^T\mathbf{y}_j[n+1] > -1 \\ 0 & \textit{if} & \mathbf{v}[n+1]^T\mathbf{y}_j[n+1] \leq -1 \end{matrix} \tag{7}$$

The penalty function $\varphi_j^+(\mathbf{v}[n+1])$ is equal to zero if and only if the vector $\mathbf{y}_j[n+1]$ ($\mathbf{y}_j[n+1] \in G^+$) is situated on the positive side of the hyperplane $H(\mathbf{v}[n+1])$ (3) and is not too near to it (Fig. 2). Similarly, $\varphi_j^-(\mathbf{v}[n+1])$ is equal to zero if the vector $\mathbf{y}_j[n+1]$ ($\mathbf{y}_j[n+1] \in G^-$) is situated on the negative side of the hyperplane $H(\mathbf{v}[n+1])$ and is not too near to it (Fig. 3).
The perceptron criterion function $\Phi(\mathbf{v}[n+1])$ is defined on the sets G^+ and G^- (1) as the weighted sum of the penalty functions $\varphi_j^+(\mathbf{v}[n+1])$ and $\varphi_j^-(\mathbf{v}[n+1])$ (Bobrowski, 2005):

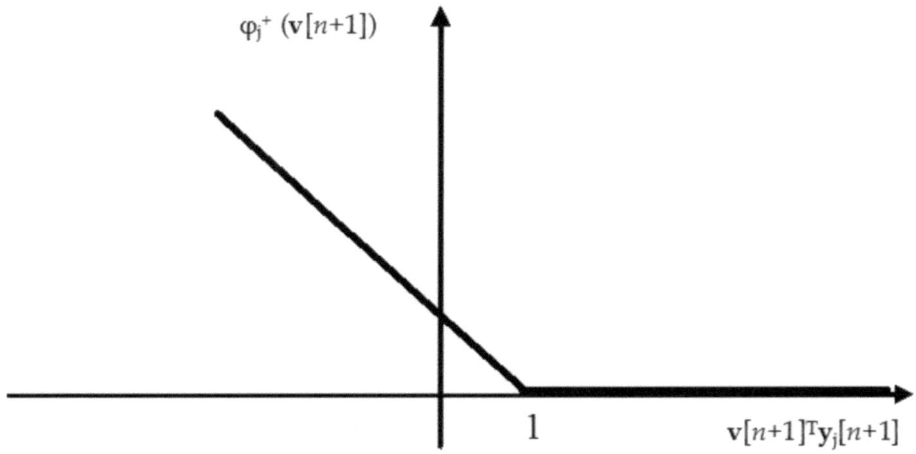

Fig. 2. The positive penalty function $\varphi_j^+(\mathbf{v}[n+1])$ (6).

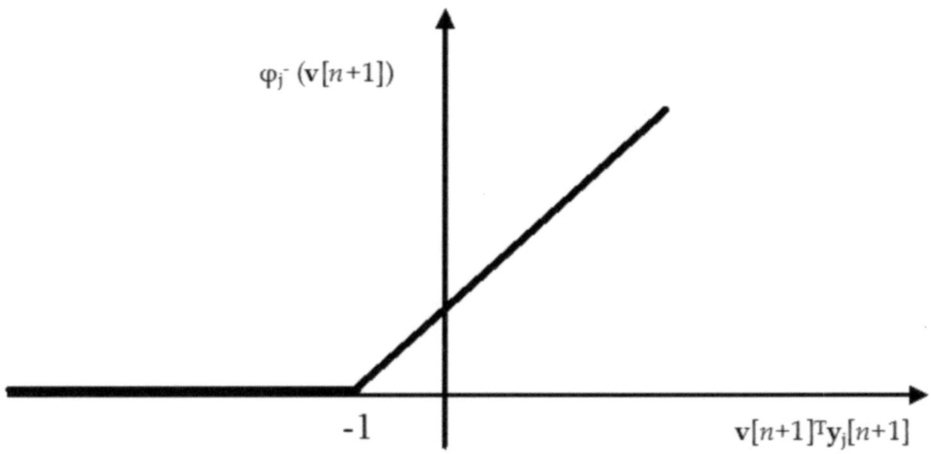

Fig. 3. The negative penalty function $\varphi_j^-(\mathbf{v}[n+1])$ (7).

$$\Phi(\mathbf{v}[n+1]) = \sum_{j \in J+} \alpha_j \varphi_j^+(\mathbf{v}[n+1]) + \sum_{j \in J-} \alpha_j \varphi_j^-(\mathbf{v}[n+1]) \quad (8)$$

where nonnegative parameters α_j determine *prices* of particular feature vectors $\mathbf{x}_j[n]$.
We are interested in the finding minimum $\Phi(\mathbf{v}_k^*[n+1])$ of the criterion function $\Phi(\mathbf{v}[n+1])$:

$$(\forall \mathbf{v}[n+1]) \; \Phi(\mathbf{v}[n+1]) \geq \Phi(\mathbf{v}_k^*[n+1]) = \Phi^* \quad (9)$$

It has been proved that the minimal value Φ^* is equal to zero ($\Phi^* = 0$) if and only if the sets G^+ and G^- (1) are linearly separable (4) (Bobrowski, 2005).

$$(\Phi^* = 0) \Leftrightarrow (G^+ \text{ and } G^- \textit{ are linearly separable } (4)) \quad (10)$$

A modified *CPL* criterion function $\Psi_\lambda(\mathbf{v}[n+1])$ which includes additional penalty functions $\phi_i(\mathbf{v}[n+1])$ and the *costs* γ_i ($\gamma_i > 0$) related to particular features x_i has been introduced in order of the feature selection (Bobrowski, 2005):

$$(\forall i \in \{1,...,n\})\ \phi_i(\mathbf{v}[n+1]) = |w_i| = \begin{cases} -\mathbf{e}_i[n+1]^T\mathbf{v}[n+1] & \text{if } \mathbf{e}_i[n+1]^T\mathbf{v}[n+1] < 0 \\ \mathbf{e}_i[n+1]^T\mathbf{v}[n+1] & \text{if } \mathbf{e}_i[n+1]^T\mathbf{v}[n+1] \geq 0 \end{cases} \tag{11}$$

and

$$\Psi_\lambda(\mathbf{v}[n+1]) = \Phi(\mathbf{v}[n+1]) + \lambda \sum_{i \in I} \gamma_i\phi_i(\mathbf{v}[n+1]) \tag{12}$$

where λ ($\lambda \geq 0$) is the *cost level*, and $I = \{1,......,n\}$.
Let us relate the hyperplane $h_j^+[n+1]$ in the parameter space R^{n+1} to each augmented feature vector $\mathbf{y}_j[n+1]$ (5) from the set G^+ (1), and the hyperplane $h_j^-[n+1]$ to each element $\mathbf{y}_j[n+1]$ (5) of the set G^-.

$$(\forall j \in J^+)\ h_j^+[n+1] = \{\mathbf{v}[n+1]: \mathbf{y}_j[n+1]^T\mathbf{v}[n+1] = 1\}$$

and (13)

$$(\forall j \in J^-)\ h_j^-[n+1] = \{\mathbf{v}[n+1]: \mathbf{y}_j[n+1]^T\mathbf{v}[n+1] = -1\}$$

The first n unit vectors $\mathbf{e}_i[n+1] = [0,...,0,1,0,...,0]^T$ ($i = 1,...,n$) without the vector $\mathbf{e}_{n+1}[n+1] = [0,\ldots,0,1]^T$ are used in defining hyperplanes $h_i^0[n+1]$ in the augmented parameter space R^{n+1} (5):

$$(\forall i \in \{1,\ldots,n\})\ h_i^0[n+1] = \{\mathbf{v}[n+1]: \mathbf{e}_i[n+1]^T\mathbf{v} = 0\} = \{\mathbf{v}[n+1]: v_i = 0\} \tag{14}$$

The hyperplanes $h_j^+[n+1]$, $h_j^-[n+1]$ and $h_i^0[n+1]$ divide the parameter space R^{n+1} (5) in the disjoined regions $R_l[n+1]$. Each region $R_l[n+1]$ is a convex polyhedron in the parameter space with number of vertices $\mathbf{v}_k[n+1]$. The *CPL* criterion function $\Psi_\lambda(\mathbf{v}[n+1])$ (12) is linear inside each region $R_l[n+1]$. It has been shown based on the theory of linear programming that the minimum of the *CPL* criterion function $\Psi_\lambda(\mathbf{v}[n+1])$ (13) can be found in one of vertices $\mathbf{v}_k[n+1]$ of some region $R_l[n+1]$ (Bobrowski, 2005). Each vertex $\mathbf{v}_k[n+1]$ in the parameter space R^{n+1} is the intersection point of at lest $(n + 1)$ hyperplanes $h_j^+[n+1]$, $h_j^-[n+1]$ or $h_i^0[n+1]$. The below equations are fulfilled in each vertex $\mathbf{v}_k[n+1]$:

$$\begin{aligned} &(\forall j \in J_k^+)\ \mathbf{y}_j[n+1]^T\mathbf{v}_k[n+1] = 1, \textit{ and} \\ &(\forall j \in J_k^-)\ \mathbf{y}_j[n+1]^T\mathbf{v}_k[n+1] = -1, \textit{ and} \\ &(\forall i \in I_k^0)\ \mathbf{e}_i[n+1]^T\mathbf{v}_k[n+1] = 0 \end{aligned} \tag{15}$$

where $J_k^+$and J_k^- are the sets of indices j such hyperplanes $h_j^+[n+1]$, $h_j^-[n+1]$ (13) that pass through the vertex $\mathbf{v}_k[n+1]$, I_k^0 is the set of indices i such hyperplanes $h_i^0[n+1]$ (14) that pass through the vertex $\mathbf{v}_k[n+1]$.
The above equations can be given in the matrix form:

$$\mathbf{B}_k[n+1]\ \mathbf{v}_k[n+1] = \delta_k'[n+1] \tag{16}$$

where $\mathbf{B}_k[n+1]$ is a non-singular matrix (*basis*) with the rows constituted by the linearly independent vectors $\mathbf{y}_j[n+1]$ ($j \in J_k^+ \cup J_k^-$) or the unit vectors $\mathbf{e}_i[n+1]$ ($i \in I_k^0$), and $\delta_k'[n+1]$ is the *margin vector* with components equal to 1, -1 or 0 according to (15).

Remark 1: The number n_1 of the independent vectors $\mathbf{y}_j[n+1]$ in the matrix $\mathbf{B}_k[n+1]$ (16) can be not greater than the *rank* r of the data set $G^+ \cup G^-$ (1). So, the number n_0 of the unit vectors $\mathbf{e}_i[n+1]$ ($i \in I_k^0$) (15) in the basis $\mathbf{B}_k[n+1]$ (16) is not less than $n - r$ ($n_0 \geq n - r$).
The vertex $\mathbf{v}_k[n+1]$ can be computed by using the basis $\mathbf{B}_k[n+1]$ and the margin vector $\delta_k'[n+1]$ (16):

$$\mathbf{v}_k[n+1] = \mathbf{B}_k[n+1]^{-1}\,\delta_k'[n+1] \tag{17}$$

The criterion function $\Psi_\lambda(\mathbf{v}[n+1])$ (12), similarly to the function $\Phi(\mathbf{v}[n+1])$ (8) is convex and piecewise-linear (*CPL*). The minimum of this function is located in one of the vertices $\mathbf{v}_k[n+1]$ (17):

$$(\exists \mathbf{v}_k\hat{}[n+1])\ (\forall \mathbf{v}[n+1])\ \Psi_\lambda(\mathbf{v}[n+1]) \geq \Psi_\lambda(\mathbf{v}_k\hat{}[n+1]) = \Psi_\lambda\hat{} \tag{18}$$

The basis exchange algorithms allow to find efficiently the optimal vertex $\mathbf{v}_k\hat{}[n+1]$ constituting the minimum of the *CPL* function $\Psi_\lambda(\mathbf{v}[n+1])$ (12), even in the case of large, multidimentional data sets G^+ and G^- (1) (Bobrowski, 1991).
Remark 2: Such components w_{ki} of the vertex $\mathbf{v}_k[n+1] = [\mathbf{w}_k[n]^T,-\theta_k]^T = [w_{k1},...,w_{kn},-\theta_k]^T$ (5) which are related to the unit vectors $\mathbf{e}_i[n+1]$ ($i \in I_k^0$) in the basis $\mathbf{B}_k[n+1]$ (16) are equal to zero ($w_{ki} = 0$) (15).
The n_0 features x_i ($i \in I_k^0$) (15) with the weights w_i equal to zero in the optimal vertex $\mathbf{v}_k\hat{}[n+1]$ (18) can be reduced without changing the separating hyperplane $H(\mathbf{w}_k\hat{}[n+1],\theta_k\hat{})$ (4). The following rule of *feature reduction* has been proposed on this base:

$$(\forall i \in I_k^0)\ \mathbf{e}_i[n+1]^T\mathbf{v}_k\hat{}[n+1] = 0 \Rightarrow w_i = 0 \Rightarrow \text{the feature } x_i \text{ is reduced} \tag{19}$$

Remark 3: A sufficiently large increase of the *cost level* λ ($\lambda \geq 0$) in the criterion function $\Psi_\lambda(\mathbf{v}[n+1])$ (12) results in an increase of the number n_0 of unit vectors $\mathbf{e}_i[n+1]$ in the basis $\mathbf{B}_k\hat{}[n+1]$ (16) linked to the optimal vertex $\mathbf{v}_k\hat{}[n+1]$ (18) (Bobrowski, 2005).
An arbitrary number n_0 of features x_i can be omitted and the feature space $F[n]$ can be reduced to the subspace $F_k\hat{}[n - n_0]$ by using of adequate value λ_k of the parameter λ in the criterion function $\Psi_\lambda(\mathbf{v}[n+1])$ (12). For example, the value $\lambda = 0$ means that the optimal vertex $\mathbf{v}_k\hat{}[n+1]$ (18) constitutes the minimum of the perceptron criterion function $\Phi(\mathbf{v}[n+1])$ (8) defined in the full feature space $F[n]$. On the other hand, sufficiently large value of the parameter λ results in the optimal vertex $\mathbf{v}_k\hat{}[n+1]$ (18) equal to zero ($\mathbf{v}_k\hat{}[n+1] = \mathbf{0}$). Such solution is not constructive, because it means that all the features x_i have been reduced (19) and the separating hyperplane $H(\mathbf{w}[n],\theta)$ (3) cannot be defined.
For a given parameter value $\lambda = \lambda_k$ (12) the optimal vertex $\mathbf{v}_k\hat{}[n+1]$ (18) is determined unambiguously as the minimum (18) of the convex and piecewise linear function $\Psi_\lambda(\mathbf{v}[n+1])$ (12). This vertex is characterized by the subset of such $n - n_0$ features x_i which are not related to the unit vectors $\mathbf{e}_i[n+1]$ ($i \notin I_k^0$) in the basis $\mathbf{B}_k\hat{}[n+1]$ (16) related to the optimal vertex $\mathbf{v}_k\hat{}[n+1]$ (18). The feature subspace $F_k\hat{}[n_k] = F_k\hat{}[n - n_0]$ can be also determined by such $n - n_0$ features x_i. Quality of the feature subspace $F_k\hat{}[n_k]$ can be determined on the basis of the quality of the optimal linear classifier designed in this subspace of dimensionality n_k. The optimal feature subspace $F_k^*[n_k]$ can be identified as one that enables create the best linear classifier. The *RLS* method of feature subset selection is based on this scheme (Bobrowski, 2008; Bobrowski & Łukaszuk, 2009).
Comparing our approach with the approach based on the least-squares criterion, we can conclude that the discriminant function based on the least-squares criterion can be linked to

the Euclidean distance L_2, whereas our method based on the convex and piece-wise linear criterion function (*CPL*) can be linked to the L_1 norm distance function.

4. Characteristics of the optimal vertices in the case of linear separabilty

Let us consider the case of "*long vectors*" in the exploratory data analysis. In this case, the dimensionality n of the feature vectors $\mathbf{x}_j[n]$ is much greater than the number m ($n >> m$) of these vectors ($j = 1,\ldots, m$). We may expect in this case that the vectors $\mathbf{x}_j[n]$ are linearly independent (Duda et al., 2001). In accordance with the *Lemma* 2, the arbitrary sets G^+ and G^- (1) of linearly independent vectors $\mathbf{x}_j[n]$ are linearly separable (6). The minimal value Φ^* (9) of the criterion function $\Phi(\mathbf{v}[n+1])$ (8) defined on linearly separable sets G^+ and G^- (1) is always equal to zero ($\Phi^* = 0$) (Bobrowski,2005). The minimum $\Phi(\mathbf{v}_k^*[n+1])$ (9) of the function $\Phi(\mathbf{v}[n+1])$ (8) can be located in the optimal vertex $\mathbf{v}_k^*[n+1]$ (9), where the below equations hold (15):

$$(\forall j \in J_k^+)\ \mathbf{v}_k^*[n']^T\mathbf{y}_j'[n'] = 1$$
$$and\ (\forall j \in J_k^-)\ \mathbf{v}_k^*[n']^T\mathbf{y}_j'[n'] = -1 \tag{20}$$

where $n' = n - n_0$ is the dimensionality of the reduced feature vectors $\mathbf{y}_j'[n']$ obtained from $\mathbf{y}_j[n+1]$ (5) after neglecting n_0 features x_i related to the set I_k^0 (15) and $\mathbf{v}_k^*[n']$ is the reduced vertex obtained from $\mathbf{v}_k^*[n+1]$ (9) by neglecting n_0 components w_i equal to zero ($w_i = 0$).
The vectors $\mathbf{y}_j'[n']$ belong to the reduced feature subspace $F_k[n']$ ($\mathbf{y}_j'[n'] \in F_k[n']$). We can remark that if the learning sets $G^+[n']$ and $G^-[n']$ constituted from the vectors $\mathbf{y}_j'[n']$ are linearly separable (4) in a given feature subspace $F_k[n']$, there may be more than one optimal vertex $\mathbf{v}_k^*[n']$ creating the minimum (9) of the function $\Phi_k(\mathbf{v}[n'])$ (8) ($\Phi_k(\mathbf{v}_k^*[n']) = 0$). In this case, each optimal vertex $\mathbf{v}_k^*[n']$ linearly separates (4) the sets $G^+[n']$ and $G^-[n']$ (Bobrowski, 2005):

$$(\forall \mathbf{y}_j'[n'] \in G^+[n'])\ \mathbf{v}_k^*[n']^T\mathbf{y}_j'[n'] \geq 1$$
$$and\ (\forall \mathbf{y}_j'[n'] \in G^-[n'])\ \mathbf{v}_k^*[n']^T\ \mathbf{y}_j'[n'] \leq -1 \tag{21}$$

Moreover, in the case of "*long vectors*" there may exist many such feature subspaces $F_k[n']$ of a given feature space $F[n]$ ($F_k[n'] \subset F[n]$) which can assure the linear separability (21). Therefore, a question arises which of the vertices $\mathbf{v}_k^*[n']$ constituting the minimum (9) of the perceptron function $\Phi(\mathbf{v}[n+1])$ (8) is the best one.
The answer for a such question can be given on the basis of minimization of the modified criterion function $\Psi_\lambda(\mathbf{v}[n+1])$ (12). In contrary to the perceptron criterion function $\Phi(\mathbf{v}[n+1])$ (8) the modified criterion function $\Psi_\lambda(\mathbf{v}[n+1])$ (12) has only one optimal vertex $\mathbf{v}_k^\wedge[n+1]$ (16). The vertex $\mathbf{v}_k^\wedge[n+1]$ (16) which constitutes minimum (18) of the function $\Psi_\lambda(\mathbf{v}[n+1])$ (12) is unambiguously determined and can be treated as the optimal one.
It can be proved that the modified criterion function $\Psi_\lambda(\mathbf{v}[n+1])$ (12) with a sufficiently small cost level λ ($\lambda \geq 0$), has the minimal value (18) in the same vertex $\mathbf{v}_k^*[n+1]$ (9) as the perceptron criterion function $\Phi(\mathbf{v}[n+1])$ (8) (Bobrowski, 2005):

$$(\exists\ \lambda_{max})\ (\forall \lambda \in (0, \lambda_{max}))\ (\forall \mathbf{v}[n+1])\ \Psi_\lambda(\mathbf{v}[n+1]) \geq \Psi_\lambda(\mathbf{v}_k^*[n+1]) \tag{22}$$

In other words, the replacement of the perceptron criterion function $\Phi(\mathbf{v}[n+1])$ (8) by the modified criterion function $\Psi_\lambda(\mathbf{v}[n+1])$ (12) does not necessarily mean changing the position of the minimum.
The modified criterion function $\Psi_\lambda(\mathbf{v}[n+1])$ (12) can be expressed in the following manner for such points $\mathbf{v}[n+1]$ which separate linearly (4) the sets G^+ and G^- (1):

$$\Psi_\lambda'(\mathbf{v}[n+1]) = \lambda \sum_{i \in I} \gamma_i \phi_i(\mathbf{v}[n+1]) = \lambda \sum_{i \in I} \gamma_i \, | \mathrm{v}_{ki} | \tag{23}$$

Therefore, the minimization of the criterion function $\Psi_\lambda(\mathbf{v}[n+1])$ (12) can be replaced by the minimization of the function $\Psi_\lambda'(\mathbf{v}[n+1])$ (23) under the condition that the point $\mathbf{v}[n+1]$ linearly separates (4) the sets G^+ and G^- (1).
Remark 5: If the sets G^+ and G^- (1) are linearly separable, then the vertex $\mathbf{v}_k^*[n+1]$ constituting the minimum of the function $\Psi_\lambda'(\mathbf{v}[n+1])$ (23) with equal feature costs γ_i has the lowest L_1 norm $||\mathbf{v}_k^*[n+1]||_{L1} = \Sigma_i |\mathrm{v}_{ki}|$ among all such vectors $\mathbf{v}[n+1]$ which linearly separate (4) these sets.
The *Remark* 5 points out a possible similarity between the *CPL* solution $\mathbf{v}_k^*[n+1]$ (22) and the optimal vector $\mathbf{v}^*[n+1]$ obtained in the *Support Vector Machines* (*SVM*) approach (Vapnik, 1998). But the use of the *CPL* function $\Psi_\lambda(\mathbf{v}[n+1])$ (12) also allows obtain other types of solutions $\mathbf{v}_k^*[n+1]$ (22) by another specification of feature costs γ_i and the cost level λ parameters. The modified criterion function $\Psi_\lambda(\mathbf{v}[n+1])$ (12) gives possibility to introduce different feature costs γ_i ($\gamma_i > 0$) related to particular features x_i. As a result, the outcome of feature subset selection process can be influenced by the feature costs γ_i (12).

5. Relaxed linear separability (*RLS*) approach to feature selection

The initial *feature space* $F[n]$ ($\mathbf{x}_j[n] \in F[n]$) is composed of the all n features x_i from a given set $\{x_1, \ldots x_n\}$. Feature reduction rule (19) results in appearance of the feature subspaces $F_k[n_k]$ ($F_k[n_k] \subset F[n]$ and $n_k < n$).
Successive increase of the value of the cost level λ in the criterion function $\Psi_\lambda(\mathbf{v}[n+1])$ (12) allows to reduce (19) additional features x_i and, as a result, allows generate the descended sequence of feature subspaces $F_k[n_k]$:

$$F[n] \supset F_1[n_1] \supset F_2[n_2] \supset \ldots \supset F_{k'}[n_{k'}], \text{ where } n_k > n_{k+1} \tag{24}$$

The sequence (24) of the feature subspaces $F_k[n_k]$ is generated in a deterministic manner on the basis data sets G^+ and G^- (1) in accordance with the relaxed linear separability (*RLS*) method (Bobrowski & Łukaszuk, 2009). Each step $F_k[n_k] \rightarrow F_{k+1}[n_{k+1}]$ is realized by an adequate increase $\lambda_k \rightarrow \lambda_{k+1} = \lambda_k + \Delta_k$ (where $\Delta_k > 0$) of the cost level λ in the criterion function $\Psi_\lambda(\mathbf{w}[n],\theta)$ (12).
One of the problems in applying the *RLS* method is to assess the quality characteristics of successive subspaces $F_k[n_k]$ (24). In this approach, a quality of a given subspace $F_k[n_k]$ is evaluated on the basis of the optimal linear classifier designed in this subspace. The better optimal linear classifier means the better feature subspace $F_k[n_k]$.
The feature subspace $F_k[n_k]$ can be obtained from the initial feature space $F[n]$ by reducing the n - n_k features x_i. Such reduction can be based on the optimal vertex $\mathbf{v}_k\hat{}[n+1]$ (18) with

the related basis $\mathbf{B}_k\hat{}[n+1]$ (16). The optimal vertex $\mathbf{v}_k\hat{}[n+1]$ (18) appoints the minimum of the criterion function $\Psi_\lambda(\mathbf{v}[n+1])$ (12) with the adequate value λ_k of the cost level λ.

Definition 2: The *reduced feature vectors* $\mathbf{y}_j'[n_k]$ ($\mathbf{y}_j'[n_k] \in F_k[n_k]$) are obtained from the feature vectors $\mathbf{y}_j[n+1] = [\mathbf{x}_j[n]^T,1]^T$ (5) after neglecting $n - n_k$ features x_i related to the set I_k^0 (15) of the optimal vertex $\mathbf{v}_k\hat{}[n+1]$ (18). The *reduced vertex* (parameter vector) $\mathbf{v}\hat{}[n_k]=[\mathbf{w}\hat{}[n_{k-1}]^T,-\theta\hat{}]^T$ (5) and $\mathbf{v}_k\hat{}[n_k]$ is obtained from the optimal vertex $\mathbf{v}_k\hat{}[n+1]$ (18) by neglecting of these $n - n_k$ components w_i, which are equal to zero ($w_i = 0$).

The reduced parameter vector $\mathbf{v}[n_k] = [\mathbf{w}[n_{k-1}]^T,-\theta]^T$ (5) defines the linear classifier $LC(\mathbf{v}[n_k])$ in the feature subspace $F_k[n_k]$. The linear classifier $LC(\mathbf{v}[n_k])$ can be characterized by the following decision rule:

$$\begin{aligned}&\textbf{if } \mathbf{v}[n_k]^T \mathbf{y}'[n_k] \geq 0, \textbf{ then } \mathbf{y}'[n_k] \textit{ is allocated to the category } \omega^+\\ &\textbf{if } \mathbf{v}[n_k]^T \mathbf{y}'[n_k] < 0, \textbf{ then } \mathbf{y}'[n_k] \textit{ is allocated to the category } \omega^-\end{aligned} \quad (25)$$

where $\mathbf{y}'[n_k] \in F_k[n_k]$, and the category (*class*) ω^+ is represented by elements $\mathbf{x}_j[n]$ of the learning set G^+ (1) and the category ω^- is represented by elements of the set G^-.

Definition 3: The *CPL optimal* linear classifier $LC(\mathbf{v}^*[n_k])$ is defined in the feature subspace $F_k[n_k]$ by a reduced parameter vector $\mathbf{v}^*[n_k]$ that constitutes the minimum $\Phi^* = \Phi_k(\mathbf{v}^*[n_k])$ (9) of the perceptron criterion function $\Phi_k(\mathbf{v}[n_k])$ (8).

The perceptron criterion function $\Phi_k(\mathbf{v}[n_k])$ is defined (8) on reduced feature vectors $\mathbf{y}_j'[n_k]$ ($\mathbf{y}_j'[n_k] \in F_k[n_k]$) that belong to the reduced learning set $G^+[n_k]$ or $G^-[n_k]$ (1).

$$G^+[n_k] = \{\mathbf{y}_j'[n_k]: j \in J^+\} \textbf{\textit{ and }} G^-[n_k] = \{\mathbf{y}_j'[n_k]: j \in J^-\} \quad (26)$$

Remark 6: The minimal value Φ_k^* of the criterion function $\Phi_k(\mathbf{v}[n_k])$ (8) on reduced feature vectors $\mathbf{y}_j'[n_k]$ is equal to zero ($\Phi_k^* = 0$) if and only if the sets $G^+[n_k]$ and $G^-[n_k]$ are linearly separable (4) in the feature subspace $F_k[n_k]$ (similarly as (10)) (Bobrowski, 2005).

It has been proved that, if the learning sets $G^+[n_k]$ and $G^-[n_k]$ (26) are linearly separable (4), then the decision rule (25) based on the optimal vector $\mathbf{v}_k^*[n_k]$ (9) allocates correctly all elements $\mathbf{y}_j'[n_k]$ of these learning sets (Bobrowski, 2005). It means that (21):

$$\begin{aligned}&(\forall \mathbf{y}_j'[n_k] \in G^+[n_k])\ \mathbf{v}^*[n_k]^T\mathbf{y}_j'[n_k] > 0, \textbf{\textit{ and}}\\ &(\forall \mathbf{y}_j'[n_k] \in G^-[n_k])\ \mathbf{v}^*[n_k]^T\mathbf{y}_j'[n_k] < 0\end{aligned} \quad (27)$$

If the sets $G^+[n_k]$ and $G^-[n_k]$ (26) are not linearly separable (4), then not all but only a majority of the vectors $\mathbf{y}_j'[n_k]$ fulfil the above inequalities.

According to the considerations of the previous paragraph, if the learning sets $G^+[n_k]$ and $G^-[n_k]$ (26) are linearly separable (4), then there is more than one vertex $\mathbf{v}_i^*[n_k]$ forming a minimum of the function $\Phi_k(\mathbf{v}[n_k])$ (8). To avoid such ambiguity, the criterion function $\Phi_k(\mathbf{v}[n_k])$ (8) can be replaced by the modified criterion function $\Psi_{k\lambda}(\mathbf{v}[n_k])$ (12) with the small value (22) of the parameter λ.

6. Evaluation of linear classifiers

The quality of the linear classifier $LC(\mathbf{v}^*[n_k])$ (25) can be evaluated by using the error estimator (*apparent error rate*) $e_a(\mathbf{v}^*[n_k])$ as the fraction of wrongly classified elements $\mathbf{y}_j'[n_k]$ of the learning sets $G^+[n_k]$ and $G^-[n_k]$ (26):

$$e_a(\mathbf{v}^*[n_k]) = m_e(\mathbf{v}^*[n_k]) \;/\; m \tag{28}$$

where m is the number of all elements $\mathbf{y}_j'[n_k]$ of the learning sets $G^+[n_k]$ and $G^-[n_k]$ (26) $\mathbf{x}_j[n]$, and $m_e(\mathbf{v}^*[n_k])$ is the number of elements $\mathbf{y}_j'[n_k]$ wrongly allocated by the rule (25).
The parameters $\mathbf{v}^*[n_k]$ of the linear classifier $LC(\mathbf{v}^*[n_k])$ (25) are estimated from the learning sets $G^+[n_k]$ and $G^-[n_k]$ (26) through minimization of the perceptron criterion function $\Phi_k(\mathbf{v}[n_k])$ (8) determined on elements $\mathbf{y}_j'[n_k]$ of these sets. It is known that if the same data $\mathbf{y}_j'[n_k]$ is used for classifier designing and for classifier evaluation, then the evaluation results are too optimistic (*biased*). The error rate (28) evaluated on the elements $\mathbf{y}_j'[n_k]$ of the learning sets is called the *apparent error* (*AE*). For example, if the learning sets $G^+[n_k]$ and $G^-[n_k]$ (26) are linearly separable (4), then the relation (27) holds and, as a result, the *apparent error* (28) evaluated on elements $\mathbf{y}_j'[n_k]$ is equal to zero ($e_a(\mathbf{v}^*[n_k]) = 0$). But it is observed in practice that the error rate of the classifier (25) evaluated on new vectors $\mathbf{y}'[n_k]$ is usually greater than zero.
For the purpose of the classifier's bias reducing, the cross validation procedures are applied (Lachenbruch, 1975). The term *p-fold cross validation* means that the learning sets $G^+[n_k]$ and $G^-[n_k]$ (26) have been divided into p parts G_i, where $i = 1,\ldots,p$ (for example $p = 10$). The vectors $\mathbf{y}_j'[n_k]$ contained in $p - 1$ parts G_i are used for definition of the criterion function $\Phi_k(\mathbf{v}[n_k])$ (8) and computing of the parameters $\mathbf{v}^*[n_k]$. The remaining vectors $\mathbf{y}_j'[n_k]$ are used as the *test set* (one part $G_{i'}$) for computing (evaluation) the error rate $e(\mathbf{v}^*[n_k])$ (28). Such evaluation is repeated p times, and each time different part $G_{i'}$ is used as the test set. The cross validlation procedure allows to use different vectors $\mathbf{y}_j'[n_k]$ (1) for the classifier (25) designing and evaluation (28) and as a result, to reduce the bias of the error rate estimation (28). The error rate (28) estimated during the *cross validation* procedure will be called the *cross-validation error* (*CVE*).
The *CVE error rate* $e_{CVE}(\mathbf{v}^*[n_k])$ (28) of the linear classifier (25) is used in the relaxed linear separability (*RLS*) method as a basic criterion in evaluation of particular feature subspaces $F_k[n_k]$ in the sequence (24) (Bobrowski & Łukaszuk, 2009). Feature subspace $F_k[n_k']$ that is linked to the linear classifier $LC(\mathbf{v}^*[n_k])$ (25) with the lowest *CVE error rate* $e_{CVE}(\mathbf{v}^*[n_k])$ can be considered as the optimal one in accordance with the *RLS* method of feature selection.

7. Toy example

The data set used in the experiment was generated by the authors. In the two-dimensional space seven points were selected. Four of them were assigned to the positive set G^+, three to the negative set G^-. The allocation of points to the sets G^+ and G^- were made in a way that the linear separability of sets was preserved. After that each point was extended to 10 dimensions. The values the remaining coordinates were drawn from the distribution N(0,1). Table 1 contains the complete data set. Features x_2 and x_7 constitute the coordinates of points in the initial two-dimensional space.
Previously described the *RLS* method was applied to the data set presented in Table 1. Table 2 shows a sequence of feature subsets studied by the method and values of the *apparent error* (28) and the *cross-validation error* obtained in particular subsets of features. The best subset of features designated by the method is a subset $F_k[2] = \{x_7, x_2\}$. It is characterized by the lowest value of the *cross-validation error*.

	x_1	x_2	x_3	x_4	x_5	x_6	x_7	x_8	x_9	x_{10}	subset
$\mathbf{x}_1[10]$	0,04	-0,20	0,66	0,83	0,12	-0,06	2,70	-0,37	0,04	-0,43	G^+
$\mathbf{x}_2[10]$	-1,47	0,70	1,16	-0,54	-0,15	-0,47	1,80	0,24	0,12	0,15	G^+
$\mathbf{x}_3[10]$	0,34	1,10	0,27	0,22	2,45	1,19	1,30	0,45	-1,06	-1,25	G^+
$\mathbf{x}_4[10]$	-1,44	2,60	1,23	-1,86	-0,31	1,26	-0,30	0,34	0,19	0,14	G^+
$\mathbf{x}_4[10]$	-0,48	-0,80	-0,55	-0,77	-0,13	0,41	1,10	-0,13	-0,83	-0,97	G^-
$\mathbf{x}_6[10]$	0,54	0,20	-0,53	0,90	-0,25	0,54	0,30	-0,34	-0,60	0,70	G^-
$\mathbf{x}_7[10]$	-0,06	1,20	1,65	-1,77	0,34	1,41	-0,80	-0,65	0,98	-0,27	G^-

Table 1. Feature vectors $\mathbf{x}_j[10]$ constituting the sets G^+ and G^-

Subset of features	AE	CVE
$F_k[5] = \{x_7, x_2, x_1, x_8, x_3\}$	0	0,28571
$F_k[4] = \{x_7, x_2, x_8, x_3\}$	0	0,14286
$F_k[3] = \{x_7, x_2, x_3\}$	0	0,14286
$F_k[2] = \{x_7, x_2\}$	0	0
$F_k[1] = \{x_7\}$	0,2619	0,28571

Table 2. Subsets of features evaluated by the RLS method, *apparent error rate* (AE) and *cross-validation error rate* (CVE) obtained in particular subsets of features

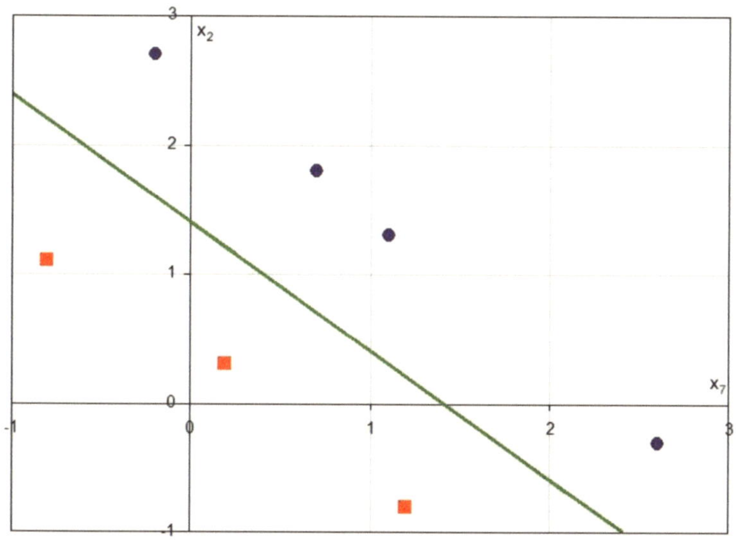

Fig. 4. Points in the feature space selected by the *RLS* method, hyperplane separated points falling within the sets G^+ (denoted by circles) and G^- (denoted by squares)

The *RLS* method in addition to the designation of the best subset of features has also determined the hyperplane separating objects from the sets G^+ and G^-.

$$H(\mathbf{w}[2],\theta) = \{\mathbf{x}[2]: 1{,}0204\ x_7 + 1{,}0884\ x_2 = 1{,}5238\} \quad (29)$$

8. Experiment on synthetic data

The data set used in the experiment contained 1000 objects, each described by 100 features. Data were drawn from a multivariate normal distribution. The values of each feature had a mean equal to 0 and standard deviation equal to 1. All the features were independent of each other (diagonal covariance matrix). The objects were divided into two disjoined subsets G^+ and G^- (1) in accordance with the values of the following linear combination:

$$3x_4+4x_{10}-7x_{17}+2x_{28}-6x_{36}+3x_{41}+3x_{58}-8x_{63}+x_{75}-x_{92}+5 \quad (30)$$

Objects corresponded to the value of expression (30) greater than 0 were assigned to subset G^+. Objects corresponded to the value of expression (30) less than 0 were assigned to subset G^-. The result was two linearly separable subsets G^+ and G^- (1) containing 630 and 370 objects.

The *RLS* method of feature selection was applied in analysis of the so-prepared synthetic data. The expected result was the preference by the method the subset of features used in the expression (30).

Figure 4 shows the *apparent error* (*AE*) and *cross-validation error* (*CVE*) values in the various tested features subspaces generated by the RLS method. Each subspace larger than 10 features ships with all 10 features used in the expression (30). Subspace of size 10 consists only of the features used in the expression (30).

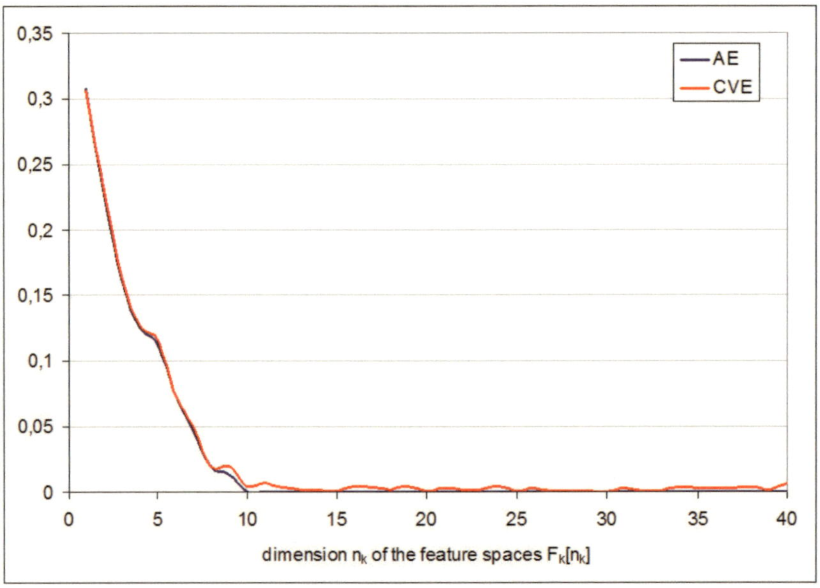

Fig. 5. The *apparent error* (AE) and the *cross-validation error* (CVE) in different feature subspaces $F_k[n_k]$ of the synthetic data set

9. Experiment on the *Leukemia* and the *Breast cancer* data sets

The *Leukemia* (Golub et al., 1999) data set contains expression levels of 7129 genes taken over 72 samples. Labels of objects indicate which of two variants of leukemia is present in the sample: acute myeloid (AML, 25 samples), or acute lymphoblastic leukemias (ALL, 47 samples).
The *Breast cancer* (van't Veer et al., 2002) data set describes the patients tested for the presence of breast cancer. The data contains 97 patient samples, 46 of which are from patients who had developed distance metastases within 5 years (labelled as "relapse"), the rest 51 samples are from patients who remained healthy from the disease after their initial diagnosis for interval of at least 5 years (labelled as "non-relapse"). The number of genes is 24481.

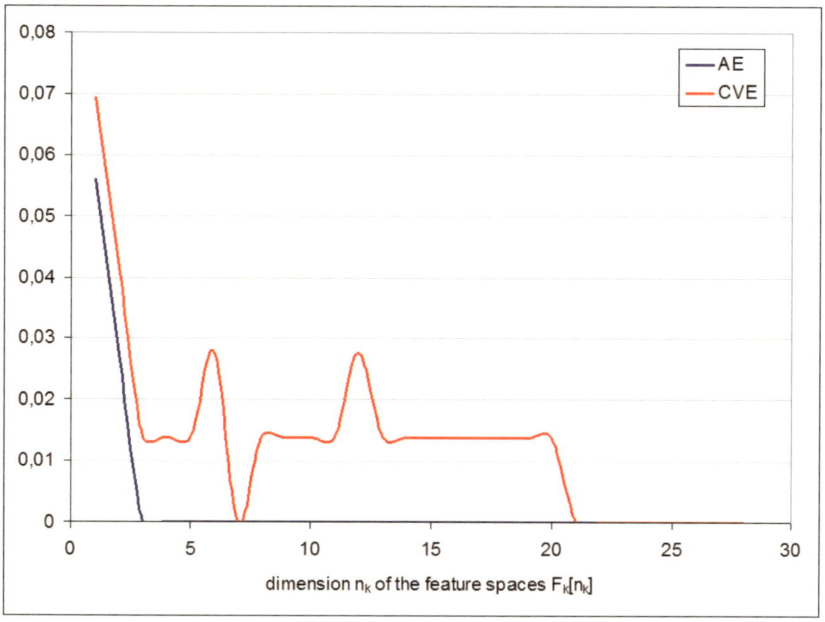

Fig. 6. The apparent error (AE) and the cross-validation error (CVE) in different feature subspaces $F_k[n_k]$ of the *Leukemia* data set

feature name	$F_k[7]$ weights w_i	$F_k[3]$ weights w_i
attribute4951	-0,99614	-1,71845
attribute1882	-0,73666	-11,6251
attribute3847	-0,55316	-
attribute6169	-0,47317	-
attribute4973	0,41573	-
attribute6539	-0,25898	-
attribute1779	-0,1519	-1,69028
threshold θ	-0,55316	2,53742

Table 3. Features x_i constituting the optimal subspace $F_k[7]$ characterised by the lowest *cross-validation error* (CVE) and features x_i constituting the lowest subspace $F_k[3]$ with *apparent error* (AE) equal to 0 of the *Leukemia* data set

Original data sets come with training and test samples that were drawn from different conditions. Here we combine them together for the purpose of cross validation. Data have also been standardized before experiment.

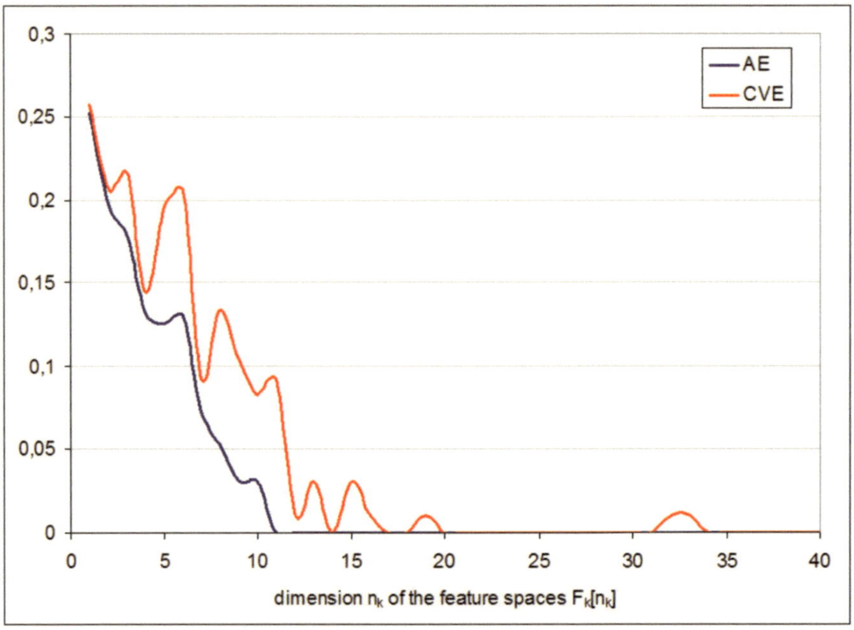

Fig. 7. The apparent error (AE) and the cross-validation error (CVE) in different feature subspaces $F_k[n_k]$ of the *Breast cancer* data set

feature name	$F_k[14]$ weights w_i	$F_k[11]$ weights w_i
Contig32002_RC	0,81467	1,89334
NM_000127	0,76305	2,67913
Contig412_RC	-0,71647	-
D86979	0,65172	-
Contig38438_RC	-0,63018	-1,58491
NM_016153	0,62345	1,81026
NM_015434	0,58631	1,43095
NM_013360	-0,58122	-1,38681
NM_002200	0,47752	1,94796
Contig44278_RC	-0,42246	-0,906564
NM_019886	0,4143	2,75393
AF055033	0,37546	1,24607
AL080059	0,31843	1,50648
NM_000909	-0,27537	-
threshold θ	0,1117	-0,247492

Table 4. Features x_i constituting the optimal subspace $F_k[14]$ characterised by the lowest *cross-validation error* (CVE) and features x_i constituting the lowest subspace $F_k[11]$ with *apparent error* (AE) equal to 0 of the *Breast cancer* data set

Figures 6 and 7 show the *apparent error* (*AE*) and *cross-validation error* (*CVE*) obtained in different feature subspaces generated by the *RLS* method. Full separability of data subsets is preserved in feature subsets much smaller than the initial very large sets of genes.

10. Conclusion

The problem of feature selection is usually resolved in practice through the evaluation of the usefulness (the validity) of individual *features* (*attributes*, *factors*) (Liu & Motoda, 2008). In this approach, resulting feature subsets are composed of such features which have the strongest individual influence on the analysed outcome. Such approach is related to the assumption about the independence of the factors. However, in a complex system, such as the living organism, these factors are often related. An advantage of the relaxed linear separability (*RLS*) method is that one may identify directly and efficiently a subset of features that influences the outcome and assesses the *combined* effect of these features as prognostic factors.

In accordance with the *RLS* method, the feature selection process involves two basic actions. The first of these actions is to generate the descending sequence (24) of feature subspaces $F_k[n_k]$. The second of the these actions is to evaluate the quality of the individual feature subspaces $F_k[n_k]$ in the sequence (24).

Generation of descending sequence (24) of feature subspaces $F_k[n_k]$ is done in the deterministic manner by multiple minimization of the criterion function the criterion function $\Psi_\lambda(\mathbf{v}[n+1])$ (12) combined with gradual increasing of the parameter λ value. The criterion function $\Psi_\lambda(\mathbf{v}[n+1])$ (12) depends on the three nonnegative parameters: α_j - *prices* of the particular feature vectors $\mathbf{x}_j[n]$ (1), γ_i - feature costs, and λ - the cost level. The composition of the consecutive feature subspaces $F_k[n_k]$ (24) depends on the choice of these parameters. For example, the costly features x_i should have a sufficiently large values of the parameter γ_i. A high value of the parameter γ_i increases the chance for elimination of a given feature x_i.

Evaluation of the quality of the individual feature subspaces $F_k[n_k]$ is based in the *RLS* method on the cross-validation of the *CPL* optimal (*Definition* 3) linear classifier (25) designed in this subspace. The optimal linear classifier (25) is designed in the feature subspace $F_k[n_k]$ through the multiple minimization of the perceptron criterion function $\Phi_k(\mathbf{v}[n_k])$ defined (8) on the reduced feature vectors $\mathbf{y}_j'[n_k]$ ($\mathbf{y}_j'[n_k] \in F_k[n_k]$) or the modified criterion function $\Psi_{k\lambda}(\mathbf{v}[n_k])$ (12) with a small value (22) of the cost level λ.

This article also contains a description of the experiments with feature selection based on the RLS method. Experiments of the first group were carried out on synthetic data. The multivariate synthetic data were generated randomly and deterministically divided into two learning sets according of predetermined key. The given key was in the form of linear combination of the unknown number of selected features. The aim of the experiment was to find an unknown key, based on available sets of multidimensional data. The experiment confirmed this possibility.

Experiments of the second group were carried out on the genetic data sets *Leucemia* (Golub et al., 1999) and *Brest cancer* (van't Veer et al., 2002). These experiments have shown, inter alia, that the *RLS* method enables finding interesting and not too large subsets of features, even if the number of features at the beginning is a huge. For example, in the case of the *Brest cancer* set, the feature space was reduced from the dimensionality n = 24481 till n_k = 11 while the linear separability (27) of the learning sets $G^+[n_k]$ and $G^-[n_k]$ (26) were preserved.

The results of calculations described in this paper were obtained by using its own implementation of the basis exchange algorithms (http://irys.wi.pb.edu.pl/dmp).

Calculations in so high dimensional feature space $F[n]$ as $n = 24481$ were made possible by achieving a high efficiency of these algorithms.

11. Acknowledgment

This work was supported by the by the NCBiR project N R13 0014 04, and partially financed by the project S/WI/2/2011 from the Białystok University of Technology, and by the project 16/St/2011 from the Institute of Biocybernetics and Biomedical Engineering PAS.

12. References

Bobrowski, L. (1991). Design of piecewise linear classifiers from formal neurons by some basis exchange technique, In: *Pattern Recognition*, 24(9), pp. 863-870

Bobrowski, L. & Łukaszuk, T. (2004). Selection of the linearly separable feature subsets, In: *Artificial Intelligence and Soft Computing - ICAISC 2004*, eds. L. Rutkowski et al., Springer Verlag, pp. 544-549

Bobrowski, L. (2005). *Eksploracja danych oparta na wypukłych i odcinkowo-liniowych funkcjach kryterialnych (Data mining based on convex and piecewise linear (CPL) criterion functions)* (in Polish), Białystok University of Technology

Bobrowski, L. (2008). Feature subsets selection based on linear separbilty, In: *Lecture Notes of the VII-th ICB Seminar: Statistics and Clinical Practice*, ed. by H. Bacelar-Nicolau, L. Bobrowski, J. Doroszewski, C. Kulikowski, N. Victor, June 2008, Warsaw

Bobrowski L. & Łukaszuk T. (2009). Feature selection based on relaxed linear separability, In: *Biocybernetical and Biomedical Engineering*, vol.29, nr 2, pp. 43-58

Duda, O. R.; Hart, P. E. & Stork D. G. (2001). *Pattern Classification*, J. Wiley, New York

Fukunaga, K. (1972). *Introduction to Statistical Pattern Recognition*, Academic Press

Golub, T. R., et al. (1999). Molecular classification of cancer: class discovery and class prediction by gene expression monitoring, *Sciences*, 286, pp. 531–537

Guyon, I.; Weston, J.; Barnhill, S. & Vapnik, V. N. (2002). Gene Selection for Cancer Classification using Support Vector Machines, In: *Machine Learning*, 46, pp. 389-422

Lachenbruch, P.A. (1975). *Discriminant Analysis*, Hafner Press, New York.

Liu, H. & Motoda, H. (Eds.) (2008). *Computational Methods of Feature Selection*, Chapmann & Hall/CRC, New York

Vapnik, V. N. (1998). *Statistical Learning Theory*, J. Wiley, New York

van't Veer, L. J., et al. (2002). Gene expression profiling predicts clinical outcome of breast cancer, *Nature*, 415(6871), pp. 530–536

7

Disease Gene Prioritization

Carlos Roberto Arias[1], Hsiang-Yuan Yeh[2] and Von-Wun Soo[1,2]
[1]*Institute of Information Systems and Applications, National Tsing Hua University*
[2]*Computer Science Department, National Tsing Hua University*
Taiwan, ROC

1. Introduction

The identification of genes is an ongoing research issue in the biomedical and bioinformatics community. The Human Genome Project which was completed in 2003, identified approximately 20,000+ genes in the human DNA, but there are still many of these genes for which their function or role is unknown, and this accounts only for healthy DNA. Genetic diseases like Cancer, Alzheimer, Hemophilia and others, have mechanisms that we currently just started to understand. For instance, genes BRCA1 and BRCA2, famous for their role in breast cancer (Friedman et al., 1994), only account for 5% of the incidence of the aforementioned cancer (Oldenburg et al., 2007). Many questions rise: What are the rest of the mechanisms involved in this cancer type? Are there other genes involved? How? This only accounts for one type of cancer, and there are at least 177 different types according to the National Cancer Institute [1]. The straightforward method to deal with this problem is to do wet lab experiments with large samples of normal and disease tissue, to test under different conditions the reactions, and check the expression or lack of it in different genes. The complication with this method is the cost, it takes time, it requires specialized equipment, and thus the economic price tag is high. Fortunately the bioinformatics area has acquired maturity during the recent years, biological data is becoming available in different formats throughout different databases and publications are providing new insights. Thanks to these, computational methods can be developed, methods that would save time, effort and money, methods that could help biomedical researchers get clues on which genes to explore on the wet laboratory, so that time is not wasted on genes that are unlikely to contribute in a given disease.

Gene Prioritization methods can be used to find genes that were previously unknown to be related to a given disease. The general definition of gene prioritization is: Given a disease D, a candidate gene set C, and the training data T, then input all these data to the method and it will compute a score for each of the candidate genes, higher scoring genes are supposed to be the genes that are most likely related to disease D, see fig. 1. Methods can be classified according to the type of input data that the method uses, as Text and Data Mining Methods and Network Based Methods. Text and Data Mining methods use training data like genetic localisation, gene expression, phenotypic data (van Driel et al., 2003), PubMeb abstracts (Tiffin et al., 2005), spatial gene expression profiles, linkage analysis (Piro et al., 2010), gene ontology and others (Adie et al., 2005; Ashburner et al., 2000; Schlicker et al., 2010); as the name suggests this

[1] http://www.cancer.gov/cancertopics/alphalist

methods mine the genome or mine the available biomedical literature to produce the scores of the candidate genes. Network Based Methods, use biological networks (Chen et al., 2009; Morrison et al., 2005) as the back bone of the prioritization method, however, some network based methods also combine some data and text mining techniques to improve their results (Aerts et al., 2006; Hutz et al., 2008).
The purpose of this chapter is to give an introduction into the Gene Prioritization Problem. Following the introduction a section explaining Biological Networks is presented as this is necessary background to understand the network based prioritization methods. After this section, we discuss about current state of the art prioritization methods with emphasis in network based methods. Next sections discuss our own prioritization method that is a network based method with a novel microarray data integration. A discussion on Challenges and Future Research opportunities follows and finally the conclusions of this chapter along with a list of available resources for Gene Prioritization.

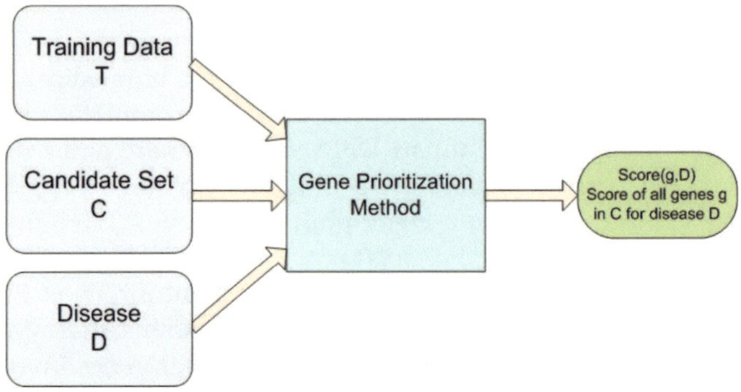

Fig. 1. General Gene Prioritization Overview

2. Biological networks

2.1 Graph theory background

A graph is a data structure that represents a set of relationships between elements or objects. Formally a graph G is a pair defined by $G = (V, E)$, where V is a set of elements that represent the nodes or vertices of the graph, the vertices in most applications hold the name of the attribute being represented. E is the set of edges, where each edge represent a relation between two vertices, an edge is defined by $E = \{(u, v)|u, v \in V\}$, this edges may hold additional information as weight, confidence or distance between nodes, therefore having $E = \{(u, v, w)|u, v \in V \text{ and } w \in Re\}$. The edges may represent direction, where $(u, v) \neq (v, u)$, in which case the graph is called directed graph, and when direction is not important, the graph is called undirected graph.

Graph Properties

Among the intrinsic properties of a graph we have: **Nodes**, the number of nodes in the network, formally $n = |V|$. **Edges**, the number of edges in the graph, formally $e = |E|$. **Connectivity** is a property of the graph, it is defined to be $connectivity(G) = \frac{e}{N}$ where $N = \binom{n}{2}$ is the maximum number of possible edges the graph can have. A graph with connectivity values closer to one would be called dense, and if the connectivity value is close to zero the graph would be called sparse, it is worth mentioning that there is no agreed exact value to consider a graph sparse or dense among the graph theory community. The

diameter of a graph is the distance of the longest shortest path on the graph. **Graph Path**, is a sequence of vertices of the form $\{v_1, v_2, v_3, ..., v_k\}$ where v_1 is the starting node and v_k is the destination node, and $(v_i, v_{i+1}) \in E$; the length of the path is defined by $l = \sum_{i=1}^{k-1} w_i$ where $w_i \in (v_i, v_{i+1}, w_i)$, when all weights are equal to 1 then the length of the path is $k - 1$. A shortest path from vertex v to u is one of the paths that has the least accumulated weight from u to v, note that there can be multiple shortest paths from one node to another.

Nodes Properties

The most basic node property is the **degree** that denotes the number of connections a node has; in directed graphs there can be a distinction between incoming and outgoing connections, called **in-degree** and **out-degree** respectively. Several measures of centrality have been created to represent how "central" a node with respect to the other nodes in the graph, this measures are: **Closeness Centrality**, based in the average shortest path to the other vertices in the network; **Betweenness Centrality**, based on the occurrence of the vertex in the shortest paths of the network, **Eigenvalue Centrality**, based in the eigenvector of the adjacency matrix that represents the graph (Freeman, 1979).

2.2 Biological networks overview

In this section a brief background on biological networks is presented. As it was explained in the previous section, a graph, or network, is a set of relationship between objects, in the specific case of biological networks those objects are related to biological processes. Typical biological networks include: gene regulation networks, signal transduction networks, metabolic networks and protein interaction networks (PIN) (Junker & Schreiber, 2008). Gene regulation networks, also known as signal transcriptional regulation networks, represent how genes control the expression of other genes; these networks are often represented by directed graphs. Signal transduction networks are an extension of gene regulation networks that represents the links between intracellular processes to extracellular functions in response to diverse external events and stimuli. The final target in a signal transduction pathway is either a transcription factor or a metabolic enzyme. Metabolic networks are determined through biochemical experiments, and consist in metabolites converting into each other with the interaction of enzymes. The last of the typical biological networks are the PINs, they represent the interaction between different gene products, they are usually modeled with undirected networks, indicating only that there is a probable functional relation between the two related proteins without indicating direction. Some other networks exist that represent specific problem oriented networks, like (Yeh et al., 2009) that identifies genetic regulatory network in prostate cancer using microarray data. Fig. 2 and Fig. 3 show the general structure of a biological network and a sample of a PIN.

2.3 Protein interaction networks

These networks are the central focus of attention in the network based disease gene prioritization, so they deserve special attention. There are four main approaches to create PIN: high throughput technology, manual curation from published experiments results, automatic text mining from published literature and computational prediction from diverse genomic data (Wu et al., 2008). Some publicly available databases hold high quality, manually curated PIN, such as HPRD (Prasad et al., 2009), BIND (Bader et al., 2003) and BioGRID (Breitkreutz et al., 2008), in our work we have used BioIR (Liu et al., 2009) which integrates the previously mentioned databases along with DIP (Salwinski et al., 2004), IntAct (Aranda et al., 2010), MIPS (Pagel et al., 2005) and MINT (Ceol et al., 2010).

Fig. 2. Structure of a Biological Network

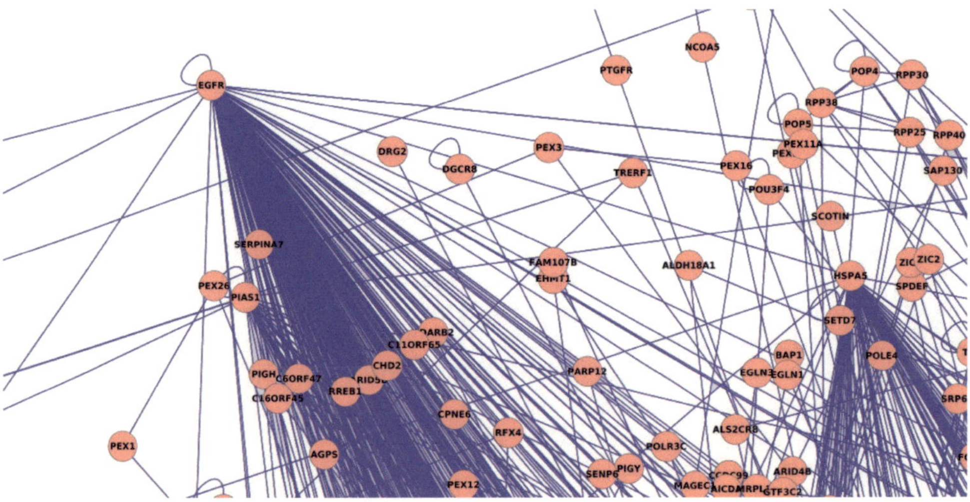

Fig. 3. Sample of Human PIN

2.3.1 Creation and curation of PIN

Current PIN databases are a rich resource of protein interactions, they mostly differ on the way they acquire their data, or on the way they validate it. For instance HPRD, BIND, BioGRID, MINT and MIPS are manually curated, this means a team of biologists check the literature to find new interactions, and once an interaction is confirmed it is added to the database. On the other hand DIP and IntAct are based on literature mining, they achieve this using computational methods that retrieve the interaction knowledge automatically from published papers. Another method to create PIN is using microarray data samples, these methods rely on the principle that co-expressing genes must be related, so by using statistical methods they can produce the list of likely relationships from the list of gene products in the microarray data.

2.3.2 Properties of PIN

PIN are known to have the following properties:

- Sparseness, although there is no one preset value of connectivity, it has been showed that biological networks are sparse containing much less than $O(n^2)$ edges in the network. Due to this property biological networks can be stored more efficiently in memory, and some algorithms exploit it to improve significantly their time performance.
- Small World, the concept was originated in the social sciences to explain how inside social networks the path length to go from one node to another is very small. However this is a subjective measure that lacks statistical or objective measure for actual networks. A more precise property that is seen in most empirical networks is Power law degree distribution, where the networks show that some few vertices have high degree and much more vertices have very low degree (Barabasi & Albert, 1999). Recent research has shown that biological networks do not necessarily follow power law degree distribution, but confirm that the distribution of degrees is heavy tailed (Garcia De Lomana et al., 2010).
- One of the disadvantageous properties of PIN is that they have a noisy nature, there is a enormous amount of missing information and false positives in the data (Edwards et al., 2002), therefore this fact must be taken into consideration when dealing with this kind of networks.
- As a consequence of the small world property, few nodes have high degree value, and it has been discovered that these nodes play an important role in the network, as opposed to other nonessential genes.
- Motifs, deep analysis in PIN has shown that there are recurrent subnetworks appearing in the full network, these subnetworks are called motifs. They have been discovered using statistical tools and showing that they occur more in the network than just by random coincidence (Junker & Schreiber, 2008).

3. Previous and on-going research

As was discussed in the introduction of the chapter prioritization methods can be classified as text and data mining based and network based methods. The main difference between the different approaches is the kind of data they use to do the prioritization of the candidate set.

3.1 Text and data mining methods

These methods usually rank candidate genes by matching their information and profile across multiple biological data sources. GeneSeeker is a web tool that selects candidate genes of the interest disease based on gene expression and phenotypic data from human and mouse (van Driel et al., 2003). eVOC system performs candidate gene selection based on the co-occurrence of disease name in PubMed abstracts through data-mining methods (Tiffin et al., 2005). DGP (Disease Gene Prediction) (López-Bigas & Ouzounis, 2004) and PROSPECTS (Adie et al., 2005) use basic sequence information to classify genes as likely or unlikely to be involved with the disease under study. The extended version of PROSPECTS, SUSPECTS (Adie et al., 2006), is developed by integrating annotation data from Gene

Ontology (GO) (Ashburner et al., 2000), InterPro and expression data. However, many of the methods suffer from limitations imposed by the data source which has little knowledge about the disease. GO terms include a brief description of the corresponding biological function of the genes but only 60% of all human genes have associated GO terms and these terms may be inconsistent due to differences in curators' judgment (Dolan et al., 2005). Due to the incomplete data, the approaches reduce the probability to rank the candidate genes of a specific disease.
Most recent methods of this kind of prioritization include MedSim (Schlicker et al., 2010) and a method based on spatially mapped gene expression (Piro et al., 2010). MedSim uses GO enrichment and applies their own similarities measures (Schlicker et al., 2007), by doing so they manage to extend the existing annotations to achieve the assignment of known disease genes to the correct phenotypes. The spatially mapped gene expression method uses a combination of data including linkage analysis, differential expression to acquire the list of candidates genes, then by using the phenotypes and associated phenotypes they find reference genes which in turn are filtered with the spatial gene-expression data; then by using both the candidate genes and the reference genes they apply their method to do the gene ranking.

3.2 Network based methods

As the name suggests these methods primarily use biological networks to do the prioritization process, this is mainly due to the increasing availability of human protein interaction data, and the emergence of network analysis. These methods usually rely on the the assumption that genes that are associated with diseases have a heavy interaction with each other (Erten & Koyutürk, 2010). Fig. 4 shows a general overview of these type of methods. The input that they commonly receive is the set of seed genes S that represent the previous knowledge to the method, genes that are known to be related to some disease D, along with these genes a score of how much they are related to the disease is given, denoted by $\sigma(v, D)$. The other part of the input is the genome of the organism represented by its PIN, denoted in the picture as the candidate set C. After the method calculates the score, just like any other method, it outputs the set of candidate genes with their score, where higher scores have higher probability of being related to disease D.

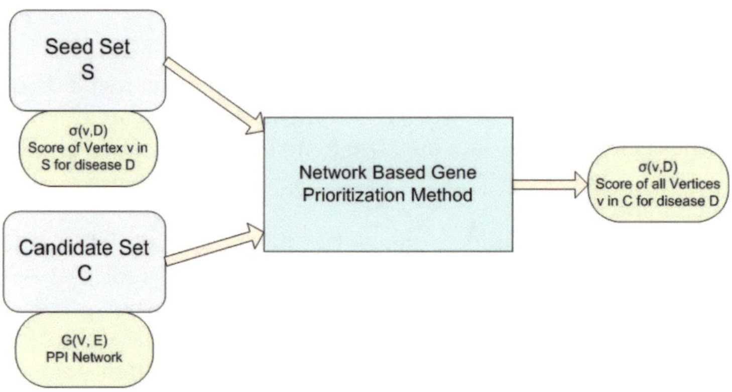

Fig. 4. Network Based Gene Prioritization Methods

Furthermore, network based methods can be classified in local and global methods. Local methods use local information to the seed genes, basically classifying by network proximity

through the inspection of the direct neighbors of the seed genes or higher order neighbors in other words nodes in the network that are not directly adjacent to the seed nodes but are easily reached by them. Global methods model the flow through the whole network to provide a score of the connectivity and impact of the seed genes or previous knowledge on the rest of the nodes.
Network based gene prioritization is performed by assessing how much genes interact together and are close to known disease genes in protein networks. Endeavour takes a machine learning approach by building a model with initial known disease-related genes as training set, then that model is used to rank the test set of candidate genes according to the similarity score using multiple genomic data sources(Aerts et al., 2006). Chen et al. applied link based strategies widely used in social and web network analyses such as HITS with priors, PageRank, and K-step Markov to prioritize disease candidate genes based on protein networks (Chen et al., 2009). Ma et al. developed a system for gene prioritization by Combining Gene expression and protein-protein Interaction network (CGI) using Markov random field theory (Ma et al., 2007). CANDID used information from publications, protein domain descriptions, cross-species conservation measures, gene expression profiles and protein-protein interactions to do a prioritization algorithm on candidate genes that influence complex human traits (Hutz et al., 2008). GeneRank ranks genes based on Google's PageRank algorithm and expression data to do gene prioritization(Morrison et al., 2005). Ozgur et al. explored the connectivity properties of biological networks to compute an association score between candidate and disease-related genes (Özgür et al., 2008). Mani et al. proposed a method called Interactome Dysregulation Enrichment Analysis (IDEA) to predict cancer related genes using interactome and microarray data (Mani et al., 2008). Karni, Soreq, and Sharan attempted to predict the causal gene from expression profile data and they identified a set of disease-related genes that could best explain the expression changes of the disease-related genes in terms of probable pathways leading from the causal to the affected genes in the network (Karni et al., 2009).Tables 1 and 2 show a summary of the aforementioned methods.

4. Gene prioritization from microarray data based on shortest paths GP-MIDAS

In this section we present our current advances in our own method: **G**ene **P**rioritization from **MI**croarray **DA**ta on **S**hortest Paths (GP-MIDAS). Our approach differs from other network based methods in the way that we assign the weights to the edges of the PIN, by doing so we manage to get considerable performance compared to other state of the art methods.

4.1 Material

We applied GP-MIDAS for the study of prostate cancer, using the following data sources:

- **PIN**: Taking advantage of the availability of public protein interaction databases, and to have a more complete protein-protein interaction network, we integrated PIN data warehouse including HPRD, DIP, BIND, IntAct, MIPS, MINT and BioGrid databases which has successfully gathered 54,283 available and non-redundant PIN pairs among 10,710 proteins into BioIR database (Liu et al., 2009).
- **Microarray Data**: We integrated microarray data taken from (Lapointe et al., 2004) that consists of 72 primary tumors and 41 normal control sample in Stanford Microarray Database (SMD) (Hubble et al., 2009).

Method	Brief Description
Gene Seeker	Gene Expression and Phenotypic Data from Human and Mouse (van Driel et al., 2003)
eVOC	Co-Occurrence of disease name on PubMed abstracts (Tiffin et al., 2005)
DGP	Basic Sequence Information (López-Bigas & Ouzounis, 2004)
PROSPECTS	Basic Sequence Information (Adie et al., 2006)
SUSPECTS	Extension of PROSPECTS, incorporates GO (Adie et al., 2005; Ashburner et al., 2000)
MedSim	Gene Ontology enrichment with their functional similarity measures (Schlicker et al., 2010)
Spatially Mapped Expression	3D Gene Expression Data, Expression Profiles, Phenotype data (Piro et al., 2010)
Limitations	Generally imposed by the data source which carries little knowledge about the disease. For instance GO terms include brief description of the corresponding biological function of the genes but only 60% of all human genes have associated GO terms, and they may be inconsistent due to differences in curators' judgement (Dolan et al., 2005).

Table 1. Data and Text Mining Gene Prioritization Methods

- **Seed Genes**: The initial seed genes known to be related to the prostate cancer are extracted from public Online Mendelian Inheritance in Man (OMIM) database which stores gene-disease associations provided by summaries of publications. The list of the seed genes are shown in Table 3.
- **Test Genes**: We took the KEGG pathway database (Kanehisa et al., 2004) and PGDB database (Li et al., 2003) that are manually curated database for prostate cancer and obtained 102 genes as the truly disease-related genes for prostate cancer. We use this set to test the accuracy of our method.

4.2 Input preparation

The collected material needs to be prepared to be useful for our method, the details on this procedure are presented as follows.

4.2.1 Cope with missing values

The microarray dataset consists of N genes and M experiments and can be represented as an $M \times N$ matrix. It presents different gene expression levels $X_{ij} \mid (i \in M, j \in N)$ in this matrix. Gene expressions either over-expressed or under-expressed can be revealed in terms of two colored channel in the microarray data representing the intensity of the cancer and normal samples, with values ranging from 0 to 255. The gene expression ratios were calculated as

Method	Brief Description	Data Sources
Endeavor	Machine Learning: Using initially known disease genes; then multiple genomic data sources to rank (Aerts et al., 2006)	BIND
HITS with Priors PageRank K-Step Markov	Prioritization Based on Networks using Social and Web Networks Analysis (Chen et al., 2009)	HPRD, BIND, BioGRID
CGI	Combination of Protein Interaction Network and Gene Expression using Markov Random Field Theory (Ma et al., 2007)	MIPS, DIP
CANDID	Uses Publications, Protein domain descriptions, cross species conservation measures, gene expression profiles and Protein Interaction Networks (Hutz et al., 2008)	NCBI Conserved Domain Database
GeneRank	Based on Google's PageRank algorithm, uses expression data (Morrison et al., 2005)	GO and Synthetic Networks
IDEA	Uses the Interactome and Microarray data (Mani et al., 2008)	B Cell Interactome and OMIM
CIPHER	Based on the assembly of a Gene-Phenotype Network (Wu et al., 2008)	HPRD and OMIM
Özgür et al. (2008)	Using connectivity properties of the networks	Literature Mining by GIN
Karni et al. (2009)	Verifies expression changes of downstream genes	HPRD
Limitations	Most of these approaches include additional interactions predicted from co-expression, pathway, functional or literature data, but still fail to incorporate weights expressing the confidence on the evidence of the interactions.	
GP-MIDAS	Our proposed method, integrates Protein Interaction Network with Normal and Disease Microarray Data, using this integration we apply all-pairs shortest paths to find the significant networks and calculate the score for the genes.	

Table 2. Network Based Gene Prioritization Methods

Gene ID	Gene Symbol	Gene name
367	AR	Androgen receptor
675	BRCA2	Breast cancer type 2 susceptibility protein
3732	CD82	CD82 antigen
11200	CHEK2	Serine/threonine-protein kinase Chk2
60528	ELAC2	Zinc phosphodiesterase ELAC protein 2
2048	EPHB2	Ephrin type-B receptor 2 precursor
3092	HIP1	Huntingtin-interacting protein 1
1316	KLF6	Krueppel-like factor 6
8379	MAD1L1	Mitotic spindle assembly checkpoint protein MAD
4481	MSR1	Macrophage scavenger receptor types I and II
4601	MXI1	MAX-interacting protein 1
7834	PCAP	Predisposing for prostate cancer
5728	PTEN/ PTENP1	Phosphatidylinositol-3,4,5-trisphosphate 3-phosphatase and dual- specificity protein phosphatase PTEN
6041	RNASEL	2-5A-dependent ribonuclease
5513	HPC1	Hereditary prostate cancer 1

Table 3. Seed Genes of Prostate Cancer from OMIM Database

the median value of the pixels minus background pixel median value for one color channel divided by the same for the other channel because the mean value of the normalized ratio is much easier to be affected by noise than the median value. We applied the base-2 logarithmic transformation of each gene among experimental dataset and this value carried out the normalization of the gene expression value with mean 0 and standard deviation 1 in every experiment. Although microarray can be used to detect thousands of genes under a variety of conditions, there are still many missing values in microarray (Troyanskaya et al., 2001). The reasons for missing values include insufficient resolution, image corruption, and dust or scratches on the slide. If a gene contains many missing values in experiments, it is not easy to determine a precise expression value for each gene that causes a difficulty in the subsequent analysis of the regulation networks. However, we can not simply remove all gene data that contains missing values because the number of remaining genes will become too small to predict the network correctly. In order to get a better result, the genes that contain less than 20% entries missing in all experiment are picked. In order to get as complete data as possible, we use the K-Nearest-Neighbors (KNN) algorithm (Troyanskaya et al., 2001) to estimate the missing values.

4.2.2 Update of microarray expression values

Once the necessary microarray data is collected, we need to preprocess it, so that it becomes ready to be used in our methods. The preprocessing procedure consists of two steps:

1. Transform the Microarray Data Expression Values. The purpose of this transformation is to make the expression values ready to be used as weights in the network. This transformation has two steps, initially the values are updated using a sample of normal expression microarray data, the effect of this operation is that values that are very similar between normal and cancer samples should have less impact on our analysis. To accomplish this we subtract the value from the cancer microarray data to the value of

the normal expression data as shown in Equation 1. The next step is to transform the values, the rationale behind this transformation is that expression values may be negative for under expressed genes, and if these values are used as they are, our network may have negative weights, thus making shortest paths analysis more difficult. Equation 2 shows how the expression values are transformed.

$$ExpressionValue_i = ExpressionValue_i - NormalExpresionValue_i \quad (1)$$

$$TransformedExpresionValue_i = -\ln\left(\frac{\mid ExpresionValue_i \mid - min}{max - min}\right) \quad (2)$$

Considering that the sign of of the value in the microarray data represents over or under expression, and the fact that we want to make a representation of distance, for this is what we want in our quantitative analysis, we use the absolute value of the microarray data, then these results are normalized, using the max and min values found, by doing these two steps we get values in the range $[0, 1]$, where values closer to 1 mean that they are more expressed (either over expressed or under expressed). Finally we compute the negative of the natural logarithm on the previous results, this is to make smaller numbers (less expression level) become large distances, and bigger numbers (higher expression level) become short distances. The result of this step is a transformation of the gene expression, where more expressed genes have smaller value, and less expressed genes have higher values, in the next step we convert this values into distances between genes, thus more expressed genes relationships will become shorter distances than less expressed genes relationships. In the case the $ExpresionValue_i == min$ we just set the whole result to be a big value, since $\ln(0)$ is not defined.

2. Convert to Human Protein Interaction Network to a Weighted Network. Since we need the network to become a weighted one, where these weights are related to the specific interactions in cancer related network, we use the transformed values of the microarray data. However the microarray data provides transformed expression values for the genes, not for the relationship between genes. The Pearson correlation coefficient for analyzing gene-pair relationships could be unsuitable to explore the true gene relationship because it is overly sensitive to the shape of an expression curve (Kim et al., 2007).To overcome this issue, we combine the values of the two interacting genes together. For instance if we have microarray values {(SEPW1, 4.097), (BRCA1, 1.395), (AKT1, 2.006), (BACH1, 2.823), (AHNAK, 3.597)} and we have the following edges in our graph {(AKT1, AHNAK), (BACH1, BRCA1), (BRCA1, AKT1)}, then the first edge weight would be the addition of the transformed expression values of each of the vertices 2.006 + 3.597 = 5.603 providing the weight of the first edge. The resulting weighted edges of this instance would be {(AKT1, AHNAK, 5.603), (BACH1, BRCA1, 4.218), (BRCA1, AKT1, 3.401)}.

4.3 Method description

Our current method is based on the analysis of the shortest paths between all the pairs of the genes on the input network.

4.3.1 Shortest paths analysis

Genes co-occurring in a particular network tend to participate together in related biological processes based on their linkage with the known disease genes (Tin et al., 2009). Our methodology is based in the the computation of all pairs shortest paths (APSP) in the network and the retrieval of these paths for posterior analysis in our experiments. The computation of APSP is carried out using our implementation of the all pairs shortest paths algorithm (Arias & Soo, 2010), which takes advantage of the topology and special characteristics of biological networks, such as sparseness and singles, nodes that have only one connection. We call our implementation KC-APSP. At this step of the process we input our prepared PIN and get as result a list of all the shortest paths between all the pair combination of the genes.

4.3.2 Scoring of genes on shortest paths

Once all the shortest paths are computed, we traverse the list of shortest paths ($PathList$), to verify if any of the seed genes are on the resulting paths, if so, these paths need to be considered for the scoring. Finally a score is computed for each gene. This analysis is done across M microarray data experiments.

4.3.3 Compute the score function

Having all the paths stored in $PathList_m$ for $m \in M$ we can compute the denominator $denom_m$, to be used in the score function using Equation 3, this is done for each microarray experiment m.

$$denom_m = \sum_{i=1}^{n} \frac{1}{l_{im}} \tag{3}$$

Where l_{im} is the length of the ith path in sample $m \in M$ for n generated and filtered by seed set paths. Once the denominator is ready, we proceed to compute the score. For each experiment m of M, and for each gene g on the network we compute the score for each gene according to Equation 4.

$$Score(Gene_{i,m}) = \sum_{Gene_{i,m} \in Path_{j,m}}^{PathList_m} \frac{\frac{1}{l_{j,m}}}{denom_m} \tag{4}$$

The motivation behind Equation 4 is that a gene that appears in more generated paths is going to achieve higher score, even higher for paths with shorter length, the highest being 1 if the gene appears in all the found paths.

5. Current results

In this section we present our current results, first we discuss the leave one out cross validation, and lastly we present the precision and recall of our method compared to other methods that use similar data sources to GP-MIDAS. As it is shown in this section, our method presents promising results that can lay the foundation for more advanced and accurate approaches.

5.1 Leave-one-out cross validation of our method

The performance of our algorithm was evaluated by leave-one-out cross validation method. In each experimental test on a known-disease gene set S, known as the Seed Set, which contains $|S|$ genes; we delete one gene g from the Seed Set thus having $S' = S - g$. We used S' set to train our prioritization model. Then, we prioritized the Candidate Gene Set to determine the rank of that deleted gene g. We got 100% to cover the deleted genes from the Candidate Gene Set and the rankness of those seed genes are listed in Table 4; LOO Score Position refers to the result of GP-MIDAS after deleting the given gene g from the seed set, Closeness Centrality

Gene	Recovered Subnetworks		All Seed Genes Subnet.
	LOO Score Position	Closeness Centrality Position	Closeness Centrality Position
AR	1	2	2
PTEN	6	23	24
BRCA2	10	36	26
EPHB2	13	46	42
HIP1	14	43	43
CHEK2	15	35	34
RNASEL	19	53	53
MXI1	20	58	58
MAD1L1	21	53	47
ELAC2	22	61	63
KLF6	26	42	40
MSR1	27	61	62
CD82	33	54	50

Table 4. Leave One Out Experiment Results

on Recovered Network refers to this centrality measure on the induced subgraph made from the seed set without the given gene g and all the shortest paths generated by this set pairs, and the last column refers to the centrality measure position on the full network for the given gene g. In order to realize the performance of gene prioritization with the weighted graph based on the gene expression, we compare the closeness centrality position in the entire PIN and sub-networks reconstructed from seed genes. From the entire network and sub-network of the seed genes using closeness properties, only 1 original seed genes rank among its top 20 ranking genes. However, we recover 8 seed genes among top 20 ranking genes. The results confirmed that PIN without any gene expression have more false positive and our method integrated gene expression is potentially able to perform better in the identification of genes associated with a given disease and should be more informative.

5.2 The precision and recall comparison with previous methods

We evaluated the performance of our algorithm in terms of overall precision versus recall when varying the rank threshold. Precision is the fraction of true gene-disease associations that ranked within the top k% in the corresponding trial of the cross validation procedure. Recall is the fraction of trials in which the disease-related genes from PGDB was recovered as one of the top k% scoring ones. We compare the performance with the following network based methods: GeneRanker, ENDEAVOUR, HITS with priors, PageRank, K-step Markov and CIPHER. In GeneRanker, we do not use 543 genes reported to be associated to prostate cancer in the literature but applied the seed as presented in the list of genes in Table 3. We set a back probability of 0.3 for PageRank with priors and HITS with priors, this value is selected because (Chen et al., 2009) express that this is the optimal value for back probability, and step size 6 for K-Step Markov method in ToppNet. Further, we reason that the use of literature evidence in this benchmark test would unfairly improve ENDEAVOUR's performance because these literatures may include direct evidence that reports the association between the gene and the disease. Neither one of these methods rank the seed genes therefore we only compare the performance with all the genes in the candidate set except the seed genes. Among the top 10 genes, we got 4 prostate cancer-related genes while applying both normal and cancer

Top K	Precision (%)				
	10	20	30	40	50
Methods					
HITS with Priors	40	25	23.3	17.5	18
K-Step Markov	20	20	20	22.5	18
PageRank	20	15	20	17.5	18
GeneRanker	30	20	20	20	18
Endeavour	10	15	20	20	20
CIPHER	10	10	10	20	20
GP-MIDAS	40	30	23.3	17.5	20

Table 5. Precision for Top K Rank Comparison Across Methods

Top K	Recall (%)				
	10	20	30	40	50
Methods					
HITS with Priors	4.5	5.6	7.9	7.9	10.1
K-Step Markov	2.2	4.5	6.7	10.1	10.1
PageRank	2.2	3.4	6.7	7.9	10.1
GeneRanker	3.4	4.5	6.7	9	10.1
Endeavour	1	2.9	5.9	7.8	14.6
CIPHER	1.1	2.2	3.4	9.0	11.2
GP-MIDAS	4.5	6.7	7.9	7.9	14.6

Table 6. Recall for Top K Rank Comparison Across Methods

samples and the performance is equal to the HITS with priors which is the highest one from the previous methods. We also get the highest precision among the top 50 ranking genes. Tables 5 and 6 denote that our method gets the highest precision and recall. Using the different expression values between cancer and normal samples may help us to extract more significant genes and rank them to be higher. Fig. 5 shows the combined precision and recall value using F-Measure, in the figure can be clearly seen that in most instances GP-MIDAS outperforms other methods.

6. Challenges and future research opportunities

As it was previously discussed gene prioritization methods can be either data and text mining based or network based, however the division line between these two approaches is less clear every day as some methods integrate both approaches and use more information to improve the accuracy of the results. Despite the increase of accuracy, the main challenge is to find novel genes that are actually involved with a given disease, genes that have not been reported before, presenting the problem of proving that the newly found genes are in fact involved with the disease. Therefore it becomes essential to present more and better biological explanations on the definition of newer approaches, by doing so biomedical researchers will have more confidence in trying the novel genes in in-vitro experiments. One clear research opportunity is presented, and it is the combination of different network based approaches, by using local and global information. The work of (Erten & Koyutürk, 2010) shows promising results, aiming at the discovery of loosely connected genes using statistical correction schemes that help overcome the preference of straightforward method for genes with high centrality values; this

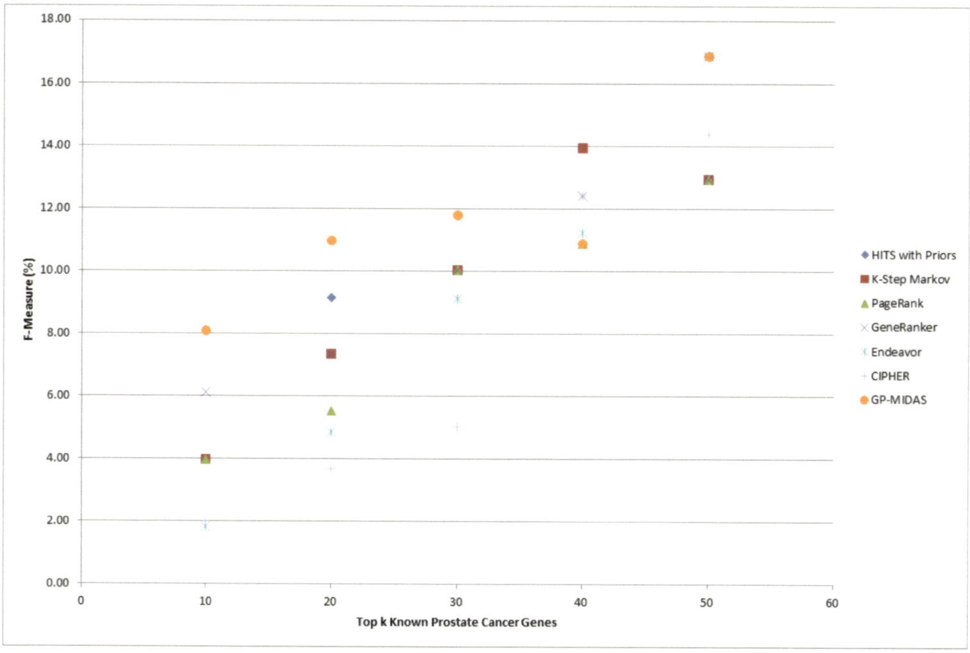

Fig. 5. F-Measure of different methods applied to Prostate Cancer

aproach uses only global methods, (Navlakha & Kingsford, 2010) combines several network based methods to produce a new score which also shows a potential new research line.

7. Conclusions

The past few years have shown an increasing interest in the disease gene prioritization problem and thanks to the availability of more and better data sources there has been a growing number of methods and approaches to this problem. A plethora of methods have become available for the genetics disease research community, and as the methods become more mature the results will become increasingly accurate and more biologically meaningful. Our own approach GP-MIDAS has proven to be promising showing a better performance in most instances to related methods, exposing that by setting the weights of the PIN to have more related meaning to the given disease the results can be better than previous plain shortest paths methodologies.

8. Acknowledgments

This research is partially supported by the Bioresources Collection and Research Center of Linko Chang Gung Memorial Hospital and National Tsing Hua University of Taiwan R. O. C. under the grant number 98N2424E1, and the Universidad Tecnológica Centroamericana in Honduras.

Resources

In this last section we present online resources, please note that since this is an evolving field, some of these resources can change with time. For a list of projects hosting biological networks see Table 7; these sites have the capability of being queried for specific proteins, or the user can also download the interaction network that is needed for his particular research. For a list of sites that offer online diseases information or software tools for disease information see Table 8. For a list of sites that offer online ontologies or software tools for ontologies see Table 9. And for a list of sites that offer online prioritization or software tools to do prioritization see Table 10.

Human Protein Reference Database	HPRD	http://www.hprd.org
Biomolecular Interaction Network Database	BIND	http://bond.unleashedinformatics.com
Biological General Repository for Interaction Datasets	BioGRID	http://thebiogrid.org/
Database of Interacting Proteins	DIP	http://dip.doe-mbi.ucla.edu/
IntAct Molecular Interaction Database	IntAct	http://www.ebi.ac.uk/intact/
The MIPS Mammalian Protein-Protein Interaction Database	MIPS	http://mips.helmholtz−muenchen.de/proj/ppi/
Molecular Interaction Database	MINT	http://mint.bio.uniroma2.it/mint/
Kyoto Encyclopedia of Genes and Genomes	KEGG	http://www.genome.jp/kegg/
National Center for Biotechnology Information	NCBI	http://www.ncbi.nlm.nih.gov/

Table 7. Available Biological Networks Sites

Online Mendelian Inheritance in Man	OMIM	http://www.ncbi.nlm.nih.gov/omim
The Human Gene Compendium	GeneCards	http://www.genecards.org/
Genetic Association Database	GAD	http://geneticassociationdb.nih.gov/
Catalog of Published Genome Wide Association Studies	GWAS	http://www.genome.gov/gwastudies/

Table 8. Available Disease Information Sites

Gene Ontology (The Gene Ontology Consortium, 2008)	http://www.geneontology.org/
eVOC Ontology (Kelso et al., 2003)	http://www.evocontology.org/
InterPro (Hunter et al., 2009)	http://www.ebi.ac.uk/interpro/

Table 9. Available Biological Ontology Sites

MedSim	http://www.funsimmat.de/
Endeavor	http://homes.esat.kuleuven.be/ bioiuser/endeavour/index.php
ToppGene	http://toppgene.cchmc.org/
Cypher	http://rulai.cshl.edu/tools/cipher/
CANDID	https://dsgweb.wustl.edu/hutz/candid.html
SUSPECTS	http://www.genetics.med.ed.ac.uk/suspects/
GP-MIDAS	http://bioir.cs.nthu.edu.tw/bne

Table 10. Available Gene Prioritization Sites

9. References

Adie, E. A., Adams, R. R., Evans, K. L., Porteous, D. J. & Pickard, B. S. (2005). Speeding disease gene discovery by sequence based candidate prioritization., *BMC Bioinformatics* 6(1). URL: *http://dx.doi.org/10.1186/1471-2105-6-55*

Adie, E. A., Adams, R. R., Evans, K. L., Porteous, D. J. & Pickard, B. S. (2006). Suspects: enabling fast and effective prioritization of positional candidates., *Bioinformatics* 22(6): 773–4. URL: *http://www.biomedsearch.com/nih/SUSPECTS-enabling-fast-effective-prioritization/16423925.html*

Aerts, S., Lambrechts, D., Maity, S., Van Loo, P., Coessens, B., De Smet, F., Tranchevent, L.-C. C., De Moor, B., Marynen, P., Hassan, B., Carmeliet, P. & Moreau, Y. (2006). Gene prioritization through genomic data fusion., *Nature biotechnology* 24(5): 537–544. URL: *http://dx.doi.org/10.1038/nbt1203*

Aranda, B., Achuthan, P., Alam-Faruque, Y., Armean, I., Bridge, A., Derow, C., Feuermann, M., Ghanbarian, A. T., Kerrien, S., Khadake, J., Kerssemakers, J., Leroy, C., Menden, M., Michaut, M., Montecchi-Palazzi, L., Neuhauser, S. N., Orchard, S., Perreau, V., Roechert, B., van Eijk, K. & Hermjakob, H. (2010). The IntAct molecular interaction database in 2010, *NUCLEIC ACIDS RESEARCH* 38(Suppl. 1): D525–D531.

Arias, C. R. & Soo, V.-W. (2010). Computing all pairs shortest paths on graphs with articulation points, *Under Review* .

Ashburner, M., Ball, C. A., Blake, J. A., Botstein, D., Butler, H., Cherry, J. M., Davis, A. P., Dolinski, K., Dwight, S. S., Eppig, J. T., Harris, M. A., Hill, D. P., Issel-Tarver, L., Kasarskis, A., Lewis, S., Matese, J. C., Richardson, J. E., Ringwald, M., Rubin, G. M. & Sherlock, G. (2000). Gene ontology: tool for the unification of biology. the gene ontology consortium., *Nature genetics* 25(1): 25–29. URL: *http://dx.doi.org/10.1038/75556*

Bader, G., Betel, D. & Hogue, C. (2003). BIND: the Biomolecular Interaction Network Database, *NUCLEIC ACIDS RESEARCH* 31(1): 248–250.

Barabasi, A. & Albert, R. (1999). Emergence of scaling in random networks, *SCIENCE* 286(5439): 509–512.

Breitkreutz, B.-J., Stark, C., Reguly, T., Boucher, L., Breitkreutz, A., Livstone, M., Oughtred, R., Lackner, D. H., Bahler, J., Wood, V., Dolinski, K. & Tyers, M. (2008). The BioGRID interaction database: 2008 update, *NUCLEIC ACIDS RESEARCH* 36(Sp. Iss. SI): D637–D640.

Ceol, A., Aryamontri, A. C., Licata, L., Peluso, D., Briganti, L., Perfetto, L., Castagnoli, L. & Cesareni, G. (2010). MINT, the molecular interaction database: 2009 update, *NUCLEIC ACIDS RESEARCH* 38(Suppl. 1): D532–D539.

Chen, J., Bardes, E. E., Aronow, B. J. & Jegga, A. G. (2009). Toppgene suite for gene list enrichment analysis and candidate gene prioritization., *Nucleic acids research* 37(Web Server issue): gkp427+.
URL: *http://dx.doi.org/10.1093/nar/gkp427*

Dolan, M. E., Ni, L., Camon, E. & Blake, J. A. (2005). A procedure for assessing go annotation consistency, *Bioinformatics* 21(suppl_1): i136–143.
URL: *http://dx.doi.org/10.1093/bioinformatics/bti1019*

Edwards, A., Kus, B., Jansen, R., Greenbaum, D., Greenblatt, J. & Gerstein, M. (2002). Bridging structural biology and genomics: assessing protein interaction data with known complexes, *TRENDS IN GENETICS* 18(10): 529–536.

Erten, S. & Koyutürk, M. (2010). Role of centrality in network-based prioritization of disease genes, *in* C. Pizzuti, M. Ritchie & M. Giacobini (eds), *Evolutionary Computation, Machine Learning and Data Mining in Bioinformatics*, Vol. 6023 of *Lecture Notes in Computer Science*, Springer Berlin / Heidelberg, pp. 13–25.

Freeman, L. (1979). Centrality in social networks: Conceptual clarification, *Social Networks* 1(3): 215–239.
URL: *http://dx.doi.org/10.1016/0378-8733(78)90021-7*

Friedman, L., Ostermeyer, E., Szabo, C., Dowd, P., Lynch, E., Rowell, S. & King, M. (1994). Confirmation of brca1 lay analysis of germline mutations linked to breast and ovarian-cancer in 10 families, *Nature genetics* 8(4): 399–404.

Garcia De Lomana, A. L., Beg, Q. K., De Fabritiis, G. & Villa-Freixa, J. (2010). Statistical Analysis of Global Connectivity and Activity Distributions in Cellular Networks, *Journal of Computational Biology* 17(7): 869–878.

Hubble, J., Demeter, J., Jin, H., Mao, M., Nitzberg, M., Reddy, T. B., Wymore, F., Zachariah, Z. K., Sherlock, G. & Ball, C. A. (2009). Implementation of genepattern within the stanford microarray database., *Nucleic acids research* 37(Database issue).
URL: *http://dx.doi.org/10.1093/nar/gkn786*

Hunter, S., Apweiler, R., Attwood, T. K., Bairoch, A., Bateman, A., Binns, D., Bork, P., Das, U., Daugherty, L., Duquenne, L., Finn, R. D., Gough, J., Haft, D., Hulo, N., Kahn, D., Kelly, E., Laugraud, A., Letunic, I., Lonsdale, D., Lopez, R., Madera, M., Maslen, J., McAnulla, C., McDowall, J., Mistry, J., Mitchell, A., Mulder, N., Natale, D., Orengo, C., Quinn, A. F., Selengut, J. D., Sigrist, C. J. A., Thimma, M., Thomas, P. D., Valentin, F., Wilson, D., Wu, C. H. & Yeats, C. (2009). InterPro: the integrative protein signature database, *NUCLEIC ACIDS RESEARCH* 37(Sp. Iss. SI): D211–D215.

Hutz, J., Kraja, A., McLeod, H. & Province, M. (2008). Candid: a flexible method for prioritizing candidate genes for complex human traits., *Genetic Epidemiology* 32(8): 779–790.
URL: *http://www.ncbi.nlm.nih.gov/pubmed/18613097*

Junker, B. H. & Schreiber, F. (eds) (2008). *Analysis of Biological Networks*, Wiley Series on Bioinformatics: Computational Techniques and Engineering, John Wiley & Sons, Inc., Hoboken, New Jersey, USA.

Kanehisa, M., Goto, S., Kawashima, S., Okuno, Y. & Hattori, M. (2004). The kegg resource for deciphering the genome., *Nucleic acids research* 32(Database issue): D277–280.
URL: *http://dx.doi.org/10.1093/nar/gkh063*

Karni, S., Soreq, H. & Sharan, R. (2009). A network-based method for predicting disease-causing genes, *Journal of Computational Biology* 16(2): 181–189.
URL: *http://dx.doi.org/10.1089/cmb.2008.05TT*

Kelso, J., Visagie, J., Theiler, G., Christoffels, A., Bardien, S., Smedley, D., Otgaar, D., Greyling, G., Jongeneel, C., McCarthy, M., Hide, T. & Hide, W. (2003). eVOC: A controlled vocabulary for unifying gene expression data, *GENOME RESEARCH* 13(6): 1222–1230.

Kim, K., Jiang, K., Zhang, S., Cai, L., beum Lee, I., Feldman, L. & Huang, H. (2007). Measuring similarities between gene expression profiles through new data transformations, *BMC Bioinformatics* 8: 29.

Lapointe, J., Li, C., Higgins, J., van de Rijn, M., Bair, E., Montgomery, K., Ferrari, M., Egevad, L., Rayford, W., Bergerheim, U., Ekman, P., DeMarzo, A., Tibshirani, R., Botstein, D., Brown, P., Brooks, J. & Pollack, J. (2004). Gene expression profiling identifies clinically relevant subtypes of prostate cancer, *Proceedings of the National Academy of Sciences of the United States of America* 101(3): 811–816.

Li, L.-C., Zhao, H., Shiina, H., Kane, C. J. & Dahiya, R. (2003). Pgdb: a curated and integrated database of genes related to the prostate, *Nucleic Acids Research* 31(1): 291–293.

Liu, H.-C., Arias, C. R. & Soo, V.-W. (2009). Bioir: An approach to public domain resource integration of human protein-protein interaction, *The proceeding of the Seventh Asia Pacific Bioinformatics Conference*.

López-Bigas, N. & Ouzounis, C. A. (2004). Genome-wide identification of genes likely to be involved in human genetic disease., *Nucleic Acids Res* 32(10): 3108–3114.
URL: *http://dx.doi.org/10.1093/nar/gkh605*

Ma, X., Lee, H., Wang, L. & Sun, F. (2007). Cgi: a new approach for prioritizing genes by combining gene expression and protein-protein interaction data, *Bioinformatics* 23(2): 215–221.
URL: *http://dx.doi.org/10.1093/bioinformatics/btl569*

Mani, K. M., Lefebvre, C., Wang, K., Lim, W. K. K., Basso, K., Dalla-Favera, R. & Califano, A. (2008). A systems biology approach to prediction of oncogenes and molecular perturbation targets in b-cell lymphomas., *Molecular systems biology* 4.
URL: *http://dx.doi.org/10.1038/msb.2008.2*

Morrison, J. L., Breitling, R., Higham, D. J. & Gilbert, D. R. (2005). Generank: using search engine technology for the analysis of microarray experiments., *BMC Bioinformatics* 6: 233. URL: *http://www.biomedsearch.com/nih/GeneRank-using-search-engine-technology/16176585.html*

Navlakha, S. & Kingsford, C. (2010). The power of protein interaction networks for associating genes with diseases, *BIOINFORMATICS* 26(8): 1057–1063.

Oldenburg, R. A., Meijers-Heijboer, H., Cornelisse, C. J. & Devilee, P. (2007). Genetic susceptibility for breast cancer: How many more genes to be found?, *CRITICAL REVIEWS IN ONCOLOGY HEMATOLOGY* 63(2): 125–149.

Özgür, A., Vu, T., Erkan, G. & Radev, D. R. (2008). Identifying gene-disease associations using centrality on a literature mined gene-interaction network, *ISMB*, pp. 277–285.

Pagel, P., Kovac, S., Oesterheld, M., Brauner, B., Dunger-Kaltenbach, I., Frishman, G., Montrone, C., Mark, P., Stumpflen, V., Mewes, H., Ruepp, A. & Frishman, D. (2005). The MIPS mammalian protein-protein interaction database, *BIOINFORMATICS* 21(6): 832–834.

Piro, R. M., Molineris, I., Ala, U., Provero, P. & Di Cunto, F. (2010). Candidate gene prioritization based on spatially mapped gene expression: an application to XLMR, *BIOINFORMATICS* 26(18): I618–I624. 9th European Conference on Computational Biology, Ghent, BELGIUM, SEP 26-29, 2010.

Prasad, T. S. K., Goel, R., Kandasamy, K., Keerthikumar, S., Kumar, S., Mathivanan, S., Telikicherla, D., Raju, R., Shafreen, B., Venugopal, A., Balakrishnan, L., Marimuthu, A., Banerjee, S., Somanathan, D. S., Sebastian, A., Rani, S., Ray, S., Kishore, C. J. H., Kanth, S., Ahmed, M., Kashyap, M. K., Mohmood, R., Ramachandra, Y. L., Krishna, V., Rahiman, B. A., Mohan, S., Ranganathan, P., Ramabadran, S., Chaerkady, R. & Pandey, A. (2009). Human Protein Reference Database-2009 update, *NUCLEIC ACIDS RESEARCH* 37(Sp. Iss. SI): D767–D772.

Salwinski, L., Miller, C., Smith, A., Pettit, F., Bowie, J. & Eisenberg, D. (2004). The Database of Interacting Proteins: 2004 update, *NUCLEIC ACIDS RESEARCH* 32(Sp. Iss. SI): D449–D451.

Schlicker, A., Lengauer, T. & Albrecht, M. (2010). Improving disease gene prioritization using the semantic similarity of Gene Ontology terms, *BIOINFORMATICS* 26(18): i561–i567. 9th European Conference on Computational Biology, Ghent, BELGIUM, SEP 26-29, 2010.

Schlicker, A., Rahnenfuehrer, J., Albrecht, M., Lengauer, T. & Domingues, F. S. (2007). GOTax: investigating biological processes and biochemical activities along the taxonomic tree, *GENOME BIOLOGY* 8(3).

The Gene Ontology Consortium (2008). The gene ontology project in 2008, *Nucl. Acids Res.* 36(36 Database issue): 440–444.
URL: *http://dx.doi.org/10.1093/nar/gkm883*

Tiffin, N., Kelso, J. F., Powell, A. R., Pan, H., Bajic, V. B. & Hide, W. A. (2005). Integration of text- and data-mining using ontologies successfully selects disease gene candidates., *Nucleic acids research* 33(5): 1544–1552.
URL: *http://dx.doi.org/10.1093/nar/gki296*

Tin, N., Andrade, M. A. & Perez-Iratxeta, C. (2009). Linking genes to diseases: it's all in the data, *Genome Medicine* 1(8): 77.

Troyanskaya, O., Cantor, M., Sherlock, G., Brown, P., Hastie, T., Tibshirani, R., Botstein, D. & Altman, R. B. (2001). Missing value estimation methods for dna microarrays., *Bioinformatics (Oxford, England)* 17(6): 520–525.
URL: *http://dx.doi.org/10.1093/bioinformatics/17.6.520*

van Driel, M., Cuelenaere, K., Kemmeren, P., Leunissen, J. & Brunner, H. (2003). A new web-based data mining tool for the identification of candidate genes for human genetic disorders, *EUROPEAN JOURNAL OF HUMAN GENETICS* 11(1): 57–63.

Wu, X., Jiang, R., Zhang, M. Q. & Li, S. (2008). Network-based global inference of human disease genes, *MOLECULAR SYSTEMS BIOLOGY* 4.

Yeh, H.-Y., Cheng, S.-W., Lin, Y.-C., Yeh, C.-Y., Lin, S.-F. & Soo, V.-W. (2009). Identifying significant genetic regulatory networks in the prostate cancer from microarray data based on transcription factor analysis and conditional independency, *BMC MEDICAL GENOMICS* 2.

8

Exploiting Protein Interaction Networks to Unravel Complex Biological Questions

Bernd Sokolowski[1] and Sandra Orchard[2]
[1]University of South Florida, College of Medicine,
[2]European Bioinformatics Institute, Wellcome Trust Genome Campus
[1]USA
[2]UK

1. Introduction

The past thirty years have witnessed a renaissance in biology as advances in technology contributed to discoveries at ever-greater orders of magnitude. One of the primary reasons for this revolution has been the advancement of technologies that allow high-throughput discovery and processing of data. This accomplishment has placed volumes of data in the realm of "discovery" science. An important point in this period came with the complete sequencing of several microbial genomes followed by the sequencing of the first multicellular organism, *Caenorhabditis elegans*, and eventually that of humans and various model organisms, such as *Drosophila melanogaster*. The edifice of the genetic code fell by wedding a biological technique developed by Sanger, known as shotgun sequencing (Sanger et al., 1977), with that of computational techniques utilizing high-speed computers. Without the advances in computer chips and processors, at a pace defined by Moore's law (Moore, 1965), sequencing would have been dramatically slower and would not have brought about the age of bioinformatics, a symbiosis of biological data, large amounts of information, and computer science.

The hypothesis that gene number is related to organism complexity is quickly discarded when comparing *Homo sapiens*, which have a genome of only 3.1 billion base pairs (Olivier et al., 2001; Venter et al., 20010), to other organisms. Estimates for the marbled lungfish, *Protopterus aethiopicus*, suggest 133 billion base pairs (Pedersen, 1971), making it the largest vertebrate genome, while, to date, the lowly amoeba, *Amoeba dubia*, is estimated to have the largest genome overall at 670 billion base pairs (McGrath & Katz, 2004). However, large genomes may be a liability, as suggested in the plant world, where *Japonica paris*, which has a genome of approximately 150 billion base pairs (Pellicer et al., 2010), grows more slowly and is more sensitive to changes in the environment (Vinogradov, 2003). In vertebrates, there appears to be an inverse correlation between genome size and brain size (Andrews & Gregory, 2009), thus, complexity may lie with other factors such as epigenetics and protein interactions. While estimates of human gene numbers rest between 20,000 - 30,000 genes, these genes may encode over 500,000 proteins. Thus, the proteome of a cell can range from several thousand proteins in prokaryotes to over 10,000 in eukaryotes. These numbers are made more daunting with the realization that approximately 80% of the proteins in a cell do not stand alone, but rather form complexes with other proteins. Moreover, this complexity increases as the proteome of a cell can change under various conditions such as stress, disease, cell cycle, etc.

Unraveling the biological complexities in a cell's fate, growth, function, death, and disease has led to a number of techniques to unlock the mechanisms to these processes. Among these advances are those that produce large amounts of data and include gene and protein arrays, phage displays, yeast two-hybrid screens, and coaffinity or coimmunoprecipitation in combination with shotgun proteomics, which utilizes mass spectrometry. Innovations in techniques exploiting mass spectrometry are of particular significance, as this technology has improved increasingly over time. With these improvements, many technologies, once outside the realm of anyone but experts, are now user-friendly, opening the possibilities for utilization by many more scientists. Consequently, these changes in technology and accessibility have led to the formation of large databanks curated by individuals with an expertise in bioinformatics. A global view of one's own data relative to those published by others increases dramatically, as one begins to delve into these databases with software tools that retrieve large amounts of deposited data.

A number of methods can be used to unravel protein pathways, but the starting point is always the wet bench experiment that will reveal the complexity of gene expression or protein-protein interactions. This first step is fraught with potential limitations and pitfalls that vary depending on the technique. However, the object is to use an approach that will allow for the capture of many protein-protein interactions without the inclusion of too many artifacts. In my lab, we have used two techniques over the last 10 years. These are the yeast two-hybrid system and coimmunoprecipitation in conjunction with two-dimensional gel electrophoresis and mass spectrometry. Both of these procedures can yield large amounts of viable data, which leaves one with the option of either cherry-picking specific protein partners or opening a whole new world, by examining the data from a global perspective using bioinformatics. The latter approach, also known as systems biology, examines the newly discovered proteins in the larger context of protein networks or interactomes. These discoveries are made by using large datasets readily available online with software that will map the many interactions once mining of these libraries is completed and integrated with the experimental data. Within these maps you will find single proteins, or nodes, connected to other proteins, via lines known as edges. In other instances, you may find clusters of proteins in which, for example, a central protein acts like the hub in a wheel, forming connections to six or more other proteins. These hubs might reveal previously unknown functions of your protein, since they can have important regulatory roles in the cell (Fox et al., 2011). Moreover, they might suggest new subcellular functions, if they are localized primarily to specific cellular organelles.

In light of these challenges, this chapter will describe the use of bench experiments and computational techniques to determine and exploit protein-protein interactions and unravel their relation to protein networks and possibly newly discovered mechanisms. These descriptions will include coimmunoprecipitation, verification using reciprocal coimmunoprecipitation and RNAi, and data mining of specific databases. Their purpose is to provide a primer without going into specific details, since many have been described previously in great detail (e.g., Golemis & Adams, 2005; Harvey & Sokolowski, 2009; Kathiresan et al., 2009; Navaratnam, 2009).

2. Experimental design and use of bait proteins

The advances in molecular biology and protein chemistry have brought a myriad of techniques to the forefront to study molecule-molecule interactions as investigators seek to wed the relationship of their molecule of interest to various mechanisms, cycles and diseases. Among the techniques that have evolved for such studies are yeast two-hybrid screening and coimmunoprecipitation/coaffinity assays. The use of one or the other depends on which technique will reveal the biologically relevant answer and which might supply the most data for obtaining large numbers of proteins to map and build networks or interactomes. The yeast two-hybrid system does not need expensive hardware, such as a mass spectrometer, it can be done in a small laboratory, while providing high-throughput capability, and it can provide reasonably quick insights into potential binding sites. However, the system is used *in vitro* with cDNA, so any search is only as good as the quality of the screened cDNA, plus any validation of findings will occur *in vivo*, eventually. Coimmunoprecipitation combined with two-dimensional electrophoresis and mass spectrometry: allows you to pull the proteins directly from the source, since you are not dependent on obtaining a cDNA library; provides insights into protein complexes and post-translational modifications; provides amino acid sequences for potentially unknown proteins. However, further studies of interacting binding sites may need the yeast two-hybrid assay or other systems *in vitro*. Both systems generate false positives and negatives.

Yeast two-hybrid screening packages the protein of interest as cDNA in an engineered viral vector or plasmid, which is used to go fishing for other proteins that are all initially dressed as cDNA and in their own plasmid. The former is the bait, whereas the latter, known as a prey, consists of a known protein or a library of unknown proteins that are, again, in the form of cDNA (cDNA library), encoding fragments of protein derived from a tissue or organism of interest. In fusion with the bait or prey cDNAs are gene sequences that respectively encode a eukaryotic binding and activating domain. Both bait and prey can be mixed together in one soup, containing yeast cells that are transformed by the plasmids, so that many will now contain a bait and a prey cDNA. The cDNAs are incorporated into the cellular machinery and expressed as protein, after which the cells are plated on an agar-based medium. If an interaction occurs between the bait and prey proteins, the activating and binding domains interact to form a transcription factor that initiates a reporter gene, thereby changing the chromatic phenotype of the yeast for visualization. The prey cDNA, encoding a fragment of protein, is isolated from the yeast and sequenced to identify the protein involved in the interaction. Typically, this procedure involves many culture plates since the more plates the more likely you will capture a number of different interacting proteins of interest. One advantage of the yeast-two hybrid system is that you can get a fairly quick picture of the domains of interaction between the bait and prey proteins, since one of the fragments pulled from the interactions likely will be a binding site. If you do not begin with this technique for high-throughput analyses, you can use it on a low scale for studying site-directed mutagenesis in a relatively quick and reliable fashion. The downside for the yeast two-hybrid approach is that you will have to obtain a cDNA library from your tissue of interest and insert the fragments into the proper plasmid. This first step can be a weakness, because the screening is only as good as the library and proteins that are weakly or indirectly associated with your protein may be lost.

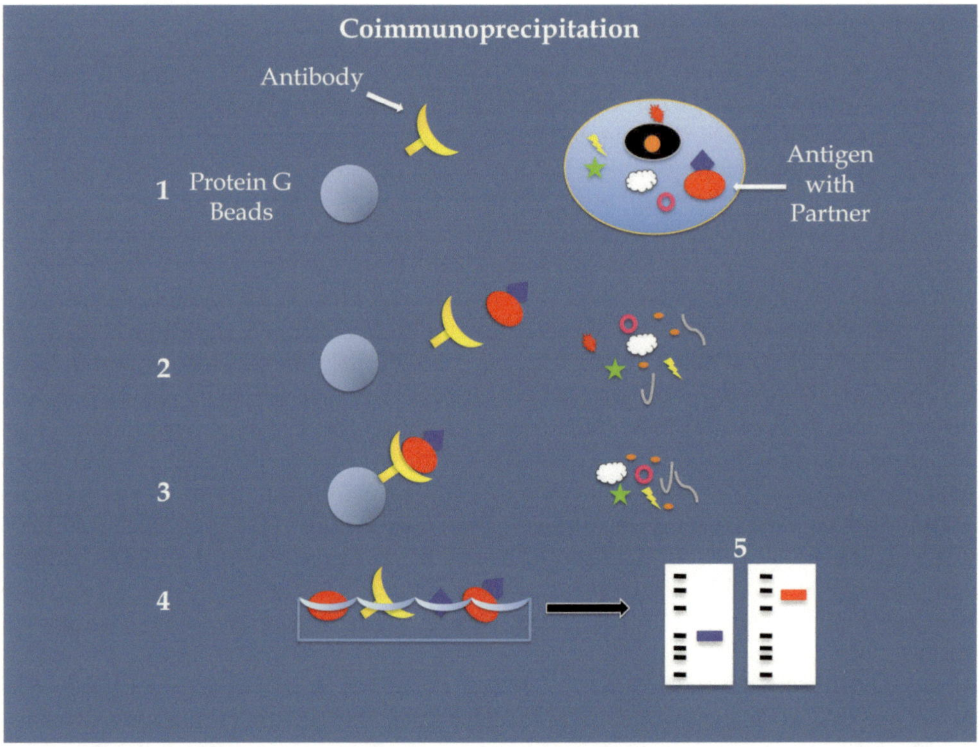

Fig. 1. Coimmunoprecipitation uses (1) a substrate consisting of protein A- or G-coated beads that bind the Fc fragment of a known antibody targeting a known antigen. (2) Once cells are homogenized the released protein lysate is mixed either with antibody alone or with antibody attached to protein-coated beads. (3) The antigen serves as bait as it brings many protein partners (prey). (4) This immunocomplex is eluted from the beads and prepared for western blotting. In high-throughput experiments, western blotting is skipped and the protein partners are separated on a 2-D gel and prepared for mass spectrometry (*see* Fig. 2). (5) For western blotting, an antibody is used to probe the blot for a known coimmunoprecipitated prey (blue). The blot can also be probed for precipitated antigen, since the known bait interacts with itself (red). In a reciprocal coimmunoprecipitation, the reverse experiment is performed, because the prey will now be used to precipitate the bait.

Coimmunoprecipitation (**Figure 1**) involves the use of an antibody on a substrate, with the antibody directed towards an antigen (the bait protein) that brings along the prey, that is, interacting proteins and protein complexes. The antibody also can be directed toward the epitope of a protein tag in fusion with a bait protein. The advantage of the technique is that you can fish in a protein soup made from your tissue or organism of interest and you can vary the antibody/antigen bait for fishing. In addition, you can pull down both direct, indirect, and weak interactions as these are highly relevant to building protein networks. The difficulty is in getting rid of interaction artifacts, so the more artifact filtering the more likely the interactions will be real. To support this effort, one can rely on a combination of centrifugation of different cellular components and using 2-D gels to better separate protein partners from one another. The components in the gel are then identified using MALDI TOF-TOF and LC-MS/MS.

3. Coimmunoprecipitation

3.1 Protein tags

The first task in setting up an experiment is to determine whether the use of cells obtained from conditions *in vivo* or *in vitro* are more suitable to the biological question at hand. At first glance, this issue may not seem relevant; however, cells obtained from conditions *in vitro* are easily accessible, allowing more freedom in the design of bait proteins and in the use of tags for quantification. A major part in this decision is determining which approach is feasible and will answer the question in a biologically relevant manner. If a system *in vitro* is chosen, there are a number of techniques that can be used, whereas the approach is more limited if cells are obtained from whole organisms or tissue lysates. For experiments *in vitro*, various heterologous expression systems are readily available, such as Chinese Hamster Ovary (CHO) and Human Embryonic Kidney (HEK) 293 cells. The increase in accessibility allows the use of isotope labeling of amino acids with Stable Isotope Labeling by Amino Acids (SILAC), Isotope-Coded Affinity Tags (ICAT), and Isobaric Tags for Relative and Absolute Quantitation (iTRAQ). These tags are successfully used in proteomic experiments involving protein-protein interactions through the differential labeling of peptides and are described extensively in a recent review (Vetter et al., 2009). Here, however, we will focus on different types of nucleotides encoding a protein tag for bait cDNA.

The cDNA of bait proteins transfected into a cell system can be epitope tagged, using FLAG (DYKDDDDK), c-myc (EQKLISEEDL), hemagglutinin (HA; YPYDVPDYA), histidine 6 (his6; HHHHHH), vesicular stomatitis virus glycoprotein (VSV-G; YTDIEMNRLGK), simian virus 5 (V5; GKPIPNPLLGLDST), and herpes simplex virus (HSV; QPELAPEDPED) tags, among others (*for additional tags see* Terpe, 2003). The FLAG tag is a hydrophobic octapeptide (Hopp et al., 1988), recognized by different anti-FLAG monoclonal antibodies (M1, M2, and M5), each with different binding and recognition characteristics. Typically this tag is used at either the N- or C-terminal ends, as is the viral hemagglutinin coat protein or HA tag. However, both can be used as an epitope tag within the C- and N-terminal domains, since tagging at the very end of either terminal may interfere with a protein-protein interaction (Duzhyy et al., 2005). Moreover, if the protein is a signaling protein, a tag at the N-terminus will be cleaved-off the main body of the protein and thus, not resolvable on a gel. These cleavage sites can be less than 20 amino acids from the N-terminus. HA tags are usually attached in multiples of two or three in fusion with a bait protein, allowing for a better signal during western blotting. The c-myc tag (Evan et al., 1985) is especially popular since there are over 150 antibodies available from different species for this particular label. In comparison, the advantage of using poly-His tags is that His binds to a chelating resin charged with metal ions such as Ni^{2+}, Cu^{2+}, or Zn^{2+} (Noronha et al., 1999; Mateo et al., 2001). It can be used to not only purify proteins, but also to bind the prey in a protein lysate poured over a bait-bound matrix in an affinity column. Once bound, the matrix-His tag can be disrupted and the prey eluted. In this scenario, lysates are used from whole organisms or tissues dissected from the organism.

3.2 Antibodies and tissue preparation

The technique for capturing protein partners is to coimmunoprecipitate protein-protein interactions using a bait antigen bound to an antibody. A second technique is to use the metal ion binding His tag in fusion with a bait protein, as mentioned above. Here, we will focus on the antibody approach, where a major hurdle is the antibody itself. These complexes can vary not only in relation to the epitope (specificity) that is targeted, but also in relation to the affinity, which can differ by source and/or fluctuate by lot number. The first rule of thumb is that that not all antibodies are created equal. Before purchasing an antibody, check that the targeted sequence of the epitope in your protein is not similar to the sequence in a different antigen. While you might assume that this comparison was made previously, particularly if the antibody is commercially available, a quick check never hurts, as sequence databases are updated on a continual basis. Gene depositories are found at the US National Institutes of Health at http://blast.ncbi.nlm.nih.gov/Blast.cgi, the European Molecular Biology Laboratory Nucleotide Sequence Database in the UK at http://www.ebi.ac.uk/Tools/sss/psiblast/, or the DNA Data Bank of Japan at http://blast.ddbj.nig.ac.jp/top-e.html. However, all three form a consortium of the International Nucleotide Sequence Database Collaboration, so information is exchanged on a daily basis. When checking, be sure to contrast species differences; however, while these differences are not fatal, the epitope should consist of 5 - 8 amino acids that are available for binding following cell/tissue denaturation. Once these sequences are checked, initial tests using western blots are valuable to determine if the antibody recognizes the denatured target.

Prior to running a coimmunoprecipitation, a necessary step is to test the chosen precipitating antibody, because many commercial antibodies are not tested for this use. Here, the second rule of thumb is that if the antibody cannot immunoprecipitate its targeted antigen (bait), it will be useless in coimmunoprecipitating any partners (prey). Thus, checking the antibody entails doing an immunoprecipitation. The procedure is similar to a coimmunoprecipitation, but rather than probing the western blot for antigen partners, you probe for the immunoprecipitated antigen. Once verified, the antibody is suitable for use in a coimmunoprecipitation. Additionally, immunoprecipitations are useful in other applications, for example, as a control with which to compare the coimmunoprecipitated species. In this scenario, you must be working with already identified proteins. As an example, use a newly discovered partner (prey), from your high-throughput experiment to coimmunoprecipitate the bait, while also immunoprecipitating the bait as control. Both co- and immunoprecipitated species should have the same weight. This step is also referred to as a reciprocal coimmunoprecipitation (discussed in section 3.2), since it validates the original bait/prey interaction. Finally, a third use for immunoprecipitation is to increase the quantity of the antigen for western blotting and Enhanced Chemiluminescent (ECL) visualization. This technique is especially useful in pulling down lowly expressed proteins. These techniques are useful for validation following the initial high-throughput experiments.

3.3 Lysate preparation and preclearing

Once an antibody is chosen and tested, preparation of the cells/tissues for coimmunoprecipitation can begin **(Figure 1)**. A step-by-step procedure was presented previously (Harvey and Sokolowski, 2009), so here, we will just touch on the salient points and limitations. The initial preparation of the tissues for coimmunoprecipitation is critical as the quality of the protein lysate is important. The goal is to disrupt the tissue sufficiently without disrupting protein-protein interactions. Thus, lysis buffers contain anywhere from 120 - 1000 mM NaCl (less to more disruptive) as well as detergents to release hydrophobic-hydrophylic interactions. Among the reagents that can disrupt protein-protein interactions are ionic detergents, such as sodium Deoxycholate (DOC) and Sodium Dodecyl Sulfate (SDS). However, nonionic detergents, such as Triton X-100, Tween20, Octyl β-D-Glucoside, N-dodecyl-β-D-Maltoside, Brij, Cymal, Digitonin, and NP-40, are useful in maintaining interactions. Octyl β-D-glucoside is especially helpful for releasing protein partners from lipid rafts, whereas n-dodecyl-β-D-maltoside isolates hydrophobic membrane proteins and preserves their activity. The isolation and separation of membrane proteins on a 2-D gel can be especially challenging. For example, our own initial studies, to cleanly separate BK channel partners from the membrane fraction on a 2-D gel, revealed amidosulfobetaine-14 (ASB-14), a zwitterionic detergent, as the best candidate relative to CHAPS (zwitterionic), octyl β-glucoside, and n-dodecyl β-D-maltopyranoside.

Once the tissue is dissected on ice and placed in a cold buffer with the proper protease and phosphatase inhibitors, any physical disruption is accomplished with pre-cooled equipment, on ice, and for short durations. These tissue perturbations include: mechanical disruption by grinding with a blade; liquid homogenization, by squeezing through a narrow space, as with a French press or Dounce; sonication, by using a vibrating probe to produce bubbles that burst and cause a sonic wave; or freeze/thaw, which bursts membranes via ice crystals. For minute tissues, such as the cochlea, use a 3 mm size probe to disrupt cells for 30 sec three times with one-minute intervals for cool down. Also, a simple mortar and pestle can be used and obtained in many different sizes. However, there is an art to the process, since you will not want to over-sonicate/homogenize. Such errors are reflected in mass spectroscopy results, where cytoplasmic proteins appear in the membrane fraction and vice versa. Again, as a reminder, the tube containing the tissues is kept on ice and any ensuing centrifugation should be done in either a refrigerator or a cold room.

Lysis buffers can be relatively standardized or they can vary from lab to lab with everyone swearing that theirs works the best. RIPA and Tris-HCl are commonly used lysis buffers and their ingredients can be easily found on the web with other types of buffers at sites such as http://www.abcam.com/index.html?pageconfig=resource&rid=11379#A1. However, some buffers contain metal chelating agents such as EGTA or EDTA. These chelators have the ability to bind or sequester metal ions, keeping them in solution and decreasing their activity. For example, EGTA sequesters Ca^{2+} and Mg^{2+}, but has a higher affinity for Ca^{2+} than Mg^{2+} ions, whereas EDTA binds Fe^{3+}, Ca^{2+}, Pb^{2+}, Co^{3+}, Mn^{2+}, and Mg^{2+}. The choice as to whether you add these chelators can depend on whether the protein-protein interactions you are interested in are metal ion dependent. The real differences come into play when deciding on which protease or phosphatase inhibitors to use (**Table 1**). Concentrations of these inhibitors can vary and may depend on, for example, whether or not you are interested in examining phosphorylated proteins. Protease/phosphatase inhibitors should always be mixed on the day of the experiment, since their stability varies quite a bit. Pepstatin A at a working solution of 1 μg/mL is stable for about one day, whereas the stock solution (100 μg/mL) is stable for several months. Leupeptin at 1-2 μg/mL is stable for a few hours, whereas the stock solution is stable for up to six months. Aprotinin, on the other hand is stable for about a week at 4°C in a solution of pH 7 at a concentration of approximately 0.5-2 μg/mL. Moreover, microcystin-LR may be preferred in place of okadaic acid, as an inhibitor of protein phosphatases PP1 and PP2A, since it is more potent. The downside of using this inhibitor is that in the U.S., microcystin is on the government list of monitored reagents and, also, it is quite expensive.

Inhibitor (Synonym)	Target
AEBSF (PEFABLOC®) ($C_8H_{10}FNO_2S$)	Serine protease
ALLN (Calpain Inhibitor I, Ac-Leu-Leu-Nle-H) ($C_{20}H_{37}N_3O_4$)	Calpain 1, cysteine proteases, cathepsin B & L
Bestatin (Ubenimex) ($C_{12}H_{24}N_2O_4$)	Aminopeptidase B, leucine amino-peptidase, triaminopeptidase
Benzamidine (Benzenecarboximidamide) ($C_7H_8N_2$)	Serine proteases, trypsin-like enzymes
β-glycerophosphate (β-glycerol phosphate) ($C_3H_9Na_2O_6P$)	Serine/Threonine phosphatases
Cathepsin inhibitor I (phenylalanyl-glycyl-NHO-Bz) ($C_{26}H_{25}N_3O_6$)	Cathepsin B, cysteine protease
DFP (Diisopropyl fluorophosphate, Isoflurophate) ($C_6H_{14}FO_3P$)	Serine proteases
E-64 (trans-epoxysuccinyl-L-leucylamido-(4-guanidino)butane) ($C_{15}H_{27}N_5O_5$)	Cysteine protease, cathepsin B, cathepsin L
Na_2 **EDTA** (Ethylenediaminetetraacetic acid, Disodium edetate) ($C_{10}H_{14}N_2Na_2O_8$)	Chelator of Ca^{2+}, Mg^{2+}, Fe^{3+}, and other metal ions
Na_2 **EGTA** (Ethylene Glycol Tetraacetic Acid) ($C_{14}H_{22}N_2Na_2O_{10}$)	Chelator more specific for Ca^{2+}
Elastatinal ($C_{21}H_{36}N_8O_7$)	Elastase
EST (L-trans-epoxysuccinyl-leucylamide (3-methyl)butane-ethyl ester, E-64d, Aloxistatin) ($C_{17}H_{30}N_2O_5$)	Calpain, cysteine protease
GGACK (Glu-Gly-Arg-chloromethylketone) ($C_{14}H_{25}ClN_6O_5$)	Urokinase
Iodoacetamide (Monoiodoacetamide) (C_2H_4INO)	Cysteine peptidase
Leupeptin (N-Acetyl-L-leucyl-L-leucyl-L-argininal) ($C_{20}H_{38}N_6O_4$)	Cysteine protease
Microcystin (Fast-death factor) ($C_{49}H_{74}N_{10}O_{12}$)	Protein phosphatases type 1 (PP1) and 2A (PP2A)
NaF (Sodium Fluoride) (NaF)	Serine/Threonine and acidic phosphatases
Na_2 **Molybdate** (Molybdic acid, disodium salt) (Na_2MoO_4)	Phosphatases
Okadaic acid (Ocadaic acid, OKA) ($C_{44}H_{68}O_{13}$)	Protein phosphatases type 1 (PP1) and 2A (PP2A)
Na_3 **Orthovanadate** (Sodium pervanadate) (Na_3O_4V)	Tyrosine phosphatases
***o*-Phenanthroline** (1,10-phenanthroline) ($C_{12}H_8N_2$)	Zn-dependent metalloproteinases
PAO (Phenylarsine oxide, Oxophenylarsine) (C_6H_5AsO)	Protein-tyrosine-phosphatase
Pepstatin A (Pepstatine) ($C_{34}H_{63}N_5O_9$)	Aspartic protease endothiapepsin
Phosphoramidon ($C_{23}H_{34}N_3O_{10}P$)	Membrane metallo-endopeptidase and endothelin-converting enzyme
PMSF (phenyl methyl sulfonyl fluoride; Benzenemethanesulfonyl fluoride) ($C_7H_7FO_2S)Na_4$	Serine proteases, fatty acid synthetase
Tetrasodium Pyrophosphate (Phosphotex, TSPP) ($Na_4O_7P_2$)	Serine/Threonine phosphatases
TLCK (Tosyllysyl chloromethyl ketone) ($C_{14}H_{21}ClN_2O_3S$)	Serine proteases
TPCK (N-tosyl-L-phenylalanine chloromethyl ketone) ($C_{17}H_{18}ClNO_3S$)	Serine endopeptidases

Table 1. Protease and phosphatase inhibitors that can be used in a cocktail mixed with a lysis buffer for protein extraction.

In order to obtain the cleanest and best protein separation on your 2-D gel, a useful step is to separate the lysate into different cellular components via centrifugation and prior to preclearing. This step is practical, especially for high-throughput experiments involving mass spectrometry (Kathiresan et al., 2009; Harvey & Sokolowski, 2009). After clearing debris, nuclei, etc., separate the membrane fraction from other soluble proteins using ultracentrifugation, by spinning the sample at 100k x g for about an hour at 4°C. The pellet will contain membrane from the plasmalemma, mitochondrion, and endoplasmic reticulum, while the supernatant will contain any remaining soluble proteins. To obtain additional separation of organelles and various other cellular components, a necessary step is density gradient centrifugation (Huber et al., 2003). However, the initial separation of membrane and cytosolic components is useful for obtaining proteins that have undergone phosphorylation or any other changes resulting from a cell's response to cycle, developmental stage, drug response, environment, disease, etc.

Prior to preclearing and coimmunoprecipitation, a choice is made with regard to the type of beads to be used as the substrate for binding the antibody. These substrates include Protein A- or G-coated agarose, sepharose or magnetic beads. Proteins A and G bind immunoglobulins in the Fc regions of an antibody, thereby, leaving the Fab region free for antigen binding. Protein A, originally derived from *Staphylococcus aureus*, binds immunoglobulins from a number of species and has a strong affinity for mouse IgG2a, 2b, 3, and rabbit IgG. Protein G was originally derived from Group G streptococcus and tends to have an affinity for a greater number of immunoglobulins across a broader range of species and subclasses of IgG. Its affinity is strong for polyclonals made from cow, horse, sheep, and mouse IgG1. Also, Protein G has less affinity for albumin, thereby decreasing background and providing cleaner preparations. Protein A and G binding affinities for various species can be found at http://www.millipore.com/immunodetection /id3/affinitypurification. The question of agarose/sepharose or magnetic beads is a matter of choice, since arguments can be made for either one. Magnetic beads are smaller at 1 – 4 μm and provide more surface area per volume, fewer handling steps, faster protocol time, greater sample recovery, and less risk of bead inclusion in the sample. However, you need a magnetic separator. In the long run, there is likely not that much difference and the outcome will lie in performing the necessary pilot experiments.

Once the tissue is cleared of debris and nuclei, separated into different cellular components, and a choice of beads is made, begin the preclearing step. Preclearing with beads involves reducing the proportion of proteins that may bind non-specifically to the agarose/sepharose beads that are used in the coimmunoprecipitation. For high-throughput experiments, where western blots are not used, it is essential. However, if the endgame is a western blot and ECL, preclear if the background masks your protein species. One limitation of preclearing is that you may lose signal, which is especially disadvantageous if the expression of your protein is low. However, signal loss can be traced by saving non-bound components during the procedure. For preclearing, the lysate is mixed with a small volume of coated beads so that any contaminating elements that increase background noise are allowed to bind over time, usually over 30 min at 4°C. The resultant complex of "sticky" proteins and beads are discarded (or saved for testing signal loss) after centrifugation and the supernatant is processed for coimmunoprecipitation. Preclearing is not to be confused with a bead control. Here, the cleared lysate is mixed with beads in the absence of antibody to form a non-immunocomplex, which is then processed and fractionated on a gel. An additional

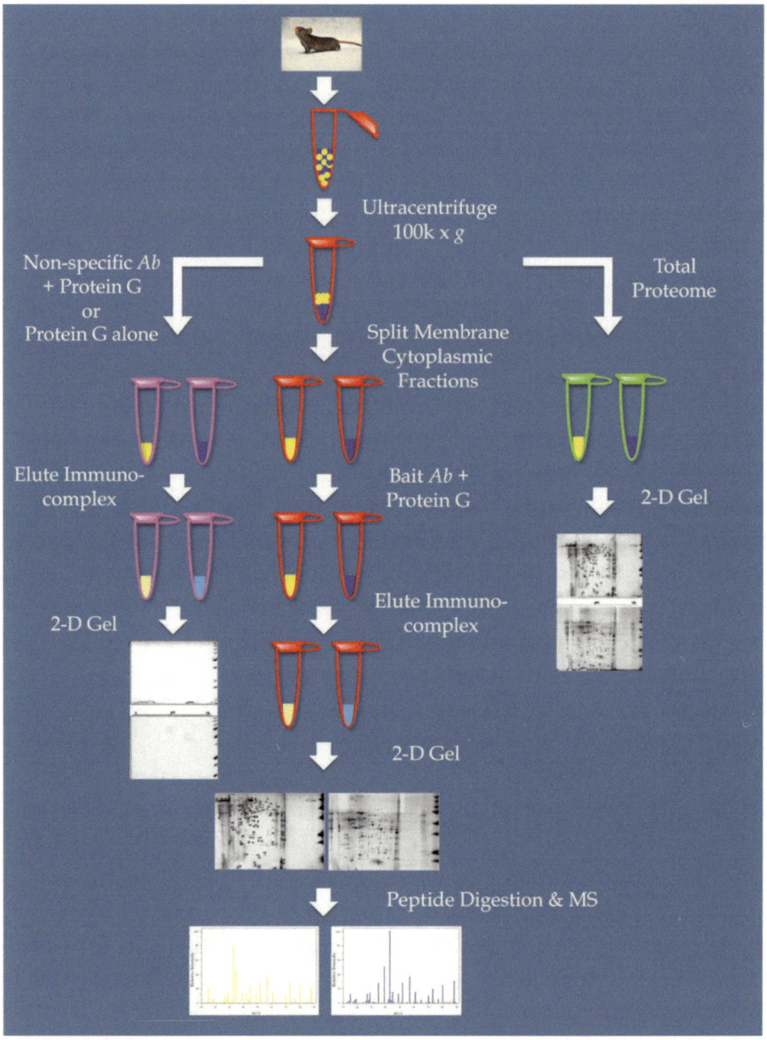

Fig. 2. Schematic of a high-throughput proteome experiment using coimmunoprecipitation, two-dimensional gels, and mass spectrometry. Initially, proteins are solubilized and separated by ultracentrifugation into membrane and cytoplasmic fractions (blue and yellow), which then are divided into two separate aliquots. Anti-bait antibody (*Ab*) with protein G beads (red tubes) is used to coimmunoprecipitate putative protein partners obtained from an organism or tissue lysate. The different subcellular fractions are probed with an antibody to a specific protein and the immunocomplex captured with Protein G-coated beads. The resultant immunocomplexes are eluted, fractionated on two-dimensional gels, and analyzed using LC-MS/MS. Control samples consist of running membrane and cytoplasmic fractions: in the absence of antibody and beads (total proteome; green tubes); with beads alone (purple tubes); or with a nonspecific antibody and beads (purple tubes).

approach to preclearing, but which can also be used as a negative control, is to use a non-specific antibody. The antibody must be isotype specific, when using a monoclonal antibody, or source specific, when using a polyclonal antibody for coimmunoprecipitation.

For example, in the event that the coimmunoprecipitating antibody is a mouse monoclonal IgG1, then use a nonspecific mouse monoclonal IgG1. If, on the other hand, the antibody to be used is a rabbit polyclonal, use a non-specific rabbit polyclonal antibody. When used as a control, mix the precleared lysate with the non-specific antibody and beads and process for gel fractionation. Finally, empirically determine how much antibody to add to the lysate fraction or to the beads, by determining the signal to noise in your result. A good starting point is to begin with 5 µg of antibody and work up or down in concentration from there.

The cells are now ready for coimmunoprecipitation for high-throughput analyses using 2-D gels (**Figure 2**). Control experiments consist of: any sticky proteins adhering to the beads (non-immunocomplexed protein), any proteins obtained using a non-specific antibody, and finally all proteins from the entire proteome of the tissue/organism. While controls may take more samples, they are of value for comparison purposes and troubleshooting, and for acceptance into high impact journals. The gel showing the total proteome is important, since it will provide an overall pattern of protein spots with which to compare the gel containing the immunocomplexed proteins. You will likely see some similar spot patterns between the gels if the separation is of a good quality. One question that will arise is whether to first bind the antibody to the beads or bind the antibody to the antigen in the lysate and then to the beads. One argument for binding the antibody to the beads first is that, since the beads are already covered with antibody, there will be a decrease in contaminant binding, and thus, a decrease in background. Antibody can be covalently bound to beads using Dimethyl Pimelimidate (DMP) Disuccinimidyl Glutarate (DSG), Disuccinimidyl Suberate (DSS), or Disuccinimidyl Tartrate (DST). Arguments against crosslinking include, the buildup of aggregates, antibodies such as monoclonals may lose their affinity, or the antibody cross-links to the beads in the incorrect position causing hindrance to antigen binding.

3.4 Fractionation and gel staining

Once both immunocomplexed and non-immunocomplexed beads are washed, perform the elution step using equal volumes of IEF sample and elution buffers, since the proteins are fractionated in two dimensions. At this point there are various nuances in terms of technique for running a 2-D gel. Among these is a step-by-step description by Kathiresan et al., (2009). Here, we will suggest some of the initial troubleshooting that may be necessary before and/or after running a full-fledged experiment. If little is known about the proteome of the tissue in your experiment, it will be of value to use an Immobilized pH Gradient (IPG) strip with a broad pH range (e.g., pH 3 - 10). Moreover, rather than initially using strips of 18 to 24 cm, which give a better resolution, use a 7 cm strip to get a quick representation of the pI ranges that you will be working with. Also, remember that the protein volume you can load is related to the size of the IPG strip, so that 7, 11, 17, 18, and 24 cm strips require volumes of 125, 185, 300, 325, and 450 µL, respectively. After separation in the first dimension, proteins are fractionated according to weight in the second dimension at which point the gel is prepared for staining.

There are several staining methods available, assuming that a CyDye was not used, since this step is accomplished prior to running the gel. The choice of stain is dependent on whether you are interested in searching for the low hanging fruit, that is, proteins that are highly expressed, or you may wish to increase staining sensitivity to detect as many protein partners as possible. Colorimetric stains include Coomassie Brilliant Blue, which will suffice for the former choice, since this stain will detect in-gel protein concentrations as low as 10 ng. For more inclusive resolution of proteins you can use silver staining, which detects protein concentrations less than 0.25 ng. For a detection range that lies between these two stains, fluorescent dyes are available that detect 0.25 - 2 ng. However, all the stains have their advantages and limitations. Coomassie Blue has less sensitivity, but is probably the most compatible stain for mass spectrometry. Silver staining has greater sensitivity but is less compatible with mass spectrometry, because, as with Coomassie Blue, the protein must be destained prior to tryptic digestion. Since formaldehyde is part of the silver staining process, cross-linking of the protein occurs (Richert et al., 2004), thereby causing problems with protein extraction from the gel and interference with mass spectrometry. A few techniques have been suggested to circumvent this problem, including ammoniacal silver staining (Richert et al., 2004; Chevallet et al., 2006). Moreover, some vendors (e.g., Thermo Scientific Pierce) optimize their reagents to make silver staining more compatible for mass spectrometry. However, silver staining still remains problematic, with its poor reproducibility and a nonlinear dynamic range, when measuring staining intensity relative to the amount of protein. There are many fluorescent stains, including those of a non-covalent variety such as SYPRO Orange, Red, Ruby, and Tangerine (Invitrogen, Carlsbad, CA, USA), ruthenium II, Deep Purple (GE Healthcare, Piscataway, NJ, USA), Krypton (Thermo Scientific, Inc., Rockford, IL, USA), and Oriole (Biorad, Hercules, CA, USA). The advantages are sensitivity, a greater dynamic range, and compatibility with mass spectrometry. Disadvantages lie primarily with cost, because of the necessity for extensive hardware for detection and quantification, the loss of signal with exposure to light, and the potential for masking certain peptides. For example, Deep Purple and SYPRO Ruby begin to lose their fluorescence after two minutes of exposure to UV transillumination, so that by 19 minutes they have lost 83% and 44% of their fluorescence, respectively (Smejkal et al., 2004). SYPRO Ruby may also inhibit identification of cysteine- and tryptophan-containing peptides (Ball & Karuso, 2007). In addition, not all fluorescent stains are compatible with the various gel types that are used to fractionate proteins for LC-MS/MS. Ruthenium II, which is much cheaper than SYPRO stains, causes increased background staining in Bis-Tris gels relative to Tris-Glycine gels (Moebius et al., 2007). These are all factors to keep in mind as part of your experimental design. Finally, once the gel is stained, any vertical streaking of protein is likely the result of insufficient equilibration or problems with the buffer solution, whereas horizontal streaking may be the result of incomplete solubilization, impurities, improper detergent, or an isoelectric focusing time that is too long or too short. With the completion of staining, the gel will need to be destained prior to removal of gel spots either manually or with a robotic arm.

4. Verification of protein partners

4.1 Manual verification of peptides

While the specifics of understanding how the data are obtained and analyzed are beyond the scope of this chapter, a quick review of some highlights are useful before describing potential experiments to validate your interactions. Once tandem mass spectrometry is completed you will obtain data derived from database search engines such as MASCOT, Seaquest, and X! Tandem. Search engines such as MASCOT generate scores as well as a compilation of spectral data. Scores above 60 can be considered valid for protein identification, assuming other parameters such as spectral data are in order. However, one may want to be conservative, especially when examining potential new partners.

Regardless, scores should not be taken at face value without analyzing the fragmentation spectra for each identified peptide, since you can have good scores and bad spectral data as well as bad scores and good spectral data. Personnel from your mass spectrometry core facility can assist in these analyses, however, you should familiarize yourself with how the ion spectra are generated and identified for your own understanding. A good starting point in comprehending spectral data is a tutorial at proteome software.com that comes in the form of a short presentation, http://www.proteomesoftware.com /Proteome_software_pro_protein_ id.html. In addition, there are many other sources of value in understanding the mechanics of mass spectrometry-based proteomics, including light reading in review articles (Aebersold and Mann, 2003), as well as more intense reading in specialized books (Gross, 2004). Also, before deciding on a core facility to analyze your precious data, assuming you have a choice, you may want to give this sobering article a quick read (Bell et al., 2009). This paper will likely push your choice towards a facility where the personnel have a great amount of experience and the search engines are continually maintained and up-to-date. Finally, the data should be analyzed for false positive rates, since some proteomics journals now require these analyses, including *Molecular and Cellular Proteomics,* which published standards in 2005 (Bradshaw, 2005).

4.2 Reciprocal coimmunoprecipitation

Once you have obtained the results, showing the putative protein-protein interactions, you can use various means to begin assessing their validity using bioinformatics and different experimental procedures. Here, we will discuss some of the methods for experimentally verifying interactions as well as assessing the potential functions of these interactions. The methods to verify are many and can depend on the protein of interest as well as the experimental question. However, one of the first and relatively easy steps for verification, considering that you've just run a two-dimensional gel, is to perform a reciprocal coimmunoprecipitation. Here, the goal is to coimmunoprecipitate the protein that was originally used as bait (antigen) by using the newfound prey (protein partner). In essence, there is a role reversal, so that the former bait is now prey and vice-versa. The means to accomplish this procedure are similar to those used for a coimmunoprecipitation using western blotting. Once you step into the realm of probing the results of a coimmunoprecipitation experiment with an antibody in a western blot, you have to consider the IgG artifacts that may appear on your film. These artifacts are the result of the presence of light (~25 kDa) and heavy (~50 kDa) immunoglobulin fragments that are detected when using an antibody from the same species for both the immunoprecipitation /coimmunoprecipitation pull-down and the western blot. The secondary antibody will recognize both IgG chains, since these are eluted from the beads along with the antibody and fractionated on the gel. The consequence is that the antigen will be masked if it has a weight similar to either IgG. Moreover, if monoclonals are used to both pull down and probe the blot, the secondary recognizes the 25 kDa band; if polyclonals are used for both, the 50 kDa band will appear as artifact. A problematic example is the use of a rabbit polyclonal antibody for both the immunoprecipitation and the western blot. To circumvent this issue, you can use antibodies from different species or use a secondary antibody consisting of HRP bound to Protein A or G. An HRP conjugated to either of these proteins will detect their non-denatured forms but not their denatured forms (Lal et al., 2005). A third solution is to use secondary antibodies that only recognize the light or heavy chain

IgGs (Jackson Immunoresearch Laboratories, Inc., Westgrove PA, USA). Thus, if your protein of interest lies in the 45 - 55 kDa range, you would use a secondary that recognizes only the light chain IgG and vice versa, if your protein lies in the 20 to 30 kDa range. However, one assumption that should not be made in running the reciprocal coimmunoprecipitation is that all the parameters in washing and stringencies are the same as for the original coimmunoprecipitation. We find that at times these variables have to be tweaked slightly differently. However, with practice these issues are usually fixed relatively quickly.

4.3 RNAi and overexpression

A method of verification that can clarify the function of your newly discovered interactions is the use of RNAi in a heterologous expression system. This approach is especially useful for proteins that lend themselves well to this sort of system, such as ion channels. There are several different types of cells to use, including HEK 293 and CHO cells, and if cell polarity is of concern, Madin-Darby Canine Kidney (MDCK) epithelial cells. These expression systems provide a vehicle for, not only expressing your proteins, but also as a means to silence the protein endogenously.

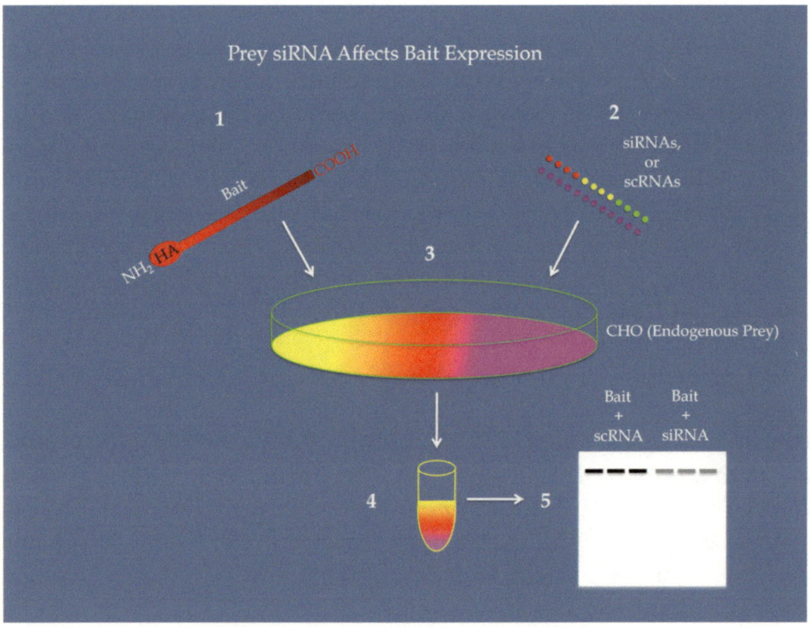

Fig. 3. Experiments, using cDNAs, are conducted to probe the function of protein-protein interactions when the prey protein is silenced in a heterologous expression system. (1) The original bait protein, in the form of an HA-tagged cDNA, is transfected with (2) siRNAs targeting specific nucleotides of the newly- discovered protein partner, or with scrambled RNAs (scRNAs; control) containing nonsense sequences. In this scenario, the (3) prey protein is endogenous to the expression system. Plates for each treatment are prepared in triplicate. (4) Cells from each treatment are scraped, homogenized and prepared for western blotting. (5) The resultant blot is probed for expression of the HA-tagged protein in both treatment and control conditions.

In order to accomplish these experiments, you will need the gene or genes that encode one or both of your proteins, that is, the bait and the prey. If the cDNA comes from either a private source or a vendor, do not assume that the construct you receive contains cDNA with a correct sequence. Be sure that it was sequenced *very* recently before use, as in, after the last amplification, because it is not uncommon to obtain constructs from either source only to discover a mutation during the course of your study, when it's too late. Once sequenced, tag the cDNA on either the C- or N-terminus with one of the specified tags discussed previously. We typically find that the antibodies to the HA and FLAG tags work quite well in these experiments. However, you have to keep in mind that the tag itself may interfere with the interaction. If the coimmunoprecipitation fails, there is no need to panic, because you can just place the tag at the other end. Also, remember that if the tag is placed on the N-terminal end of a signaling protein, you may lose your tag. Inserting the tag farther into the construct resolves this issue.

The decision then comes down to, when to silence and when to over-express, once all the necessary cDNA constructs are in order. In our work, we find that inserting and over-expressing the cDNA, encoding the original bait protein, and silencing the partner (prey) with siRNAs endogenously, works the best in our RNAi experiments (**Figure 3**). This approach entails determining that the partner, pulled from the high-throughput study, is expressed endogenously in your heterologous system. If so, check that the sequences of the siRNAs match those of the endogenous protein found in the heterologous cells. Be sure to have at least three to four siRNAs, targeting 18 - 23 bases of the sequence in different regions that are approximately 70 - 100 bases from either the 5′ or 3′ ends. Search for AA dinucleotides, since siRNAs with an overhanging UU pair at the 3′ end is the most effective, although other dinucleotides are effective to some extent (Elbashir et al., 2001; Elbashir et al., 2002). Avoid runs of G or C, since these are cut by RNAses. In addition, the GC content should not exceed 30 - 50% because the siRNA becomes less active. BLAST siRNA sequences to avoid knockdown of genes with similar sequences.

Once all materials are ready, transfect the cells with the siRNAs and the over-expressed protein (**Figure 3**). A fluorescent tag such as Cerulean can be used to check for the earliest expression of protein in live cells. We find that proteins are expressed within the first two hours, with transfection reagents such as Lipofectamine 2000 (Invitrogen, Carlsbad, CA, USA), which can be removed after four hours of transfection. However, the efficiency of transfection may vary from cell type to cell type, so a test of comparable products is needed to find the most efficient one. Anywhere from one to all of the siRNAs can be added in an equivalent ratio. Following transfection, cells are allowed to grow for approximately 48 hours at which time they are processed for protein quantification. Run triplicate plates along with a negative control, such as cells treated with scrambled RNA and over-expressed protein. Experiments are repeated a total of three times. Band densities are measured, averaged, and analyzed for statistical significance by comparing experimental versus control groups. In order to control for protein loading, perform a protein assay and verify by analyzing a control protein on the same blot as experimental and control treatments (e.g., β-actin, GAPDH).

In a similar manner, over-expression of the partner can be managed through transfection of heterologous expression systems (**Figure 4**). Again, the bait protein can be measured, but this time in response to over-expressing the prey as opposed to silencing. Transfect both constructs in a 1:1 ratio and use a control, consisting of empty vector or vector with the construct in reverse sequence along with the construct carrying the prey sequence. This procedure will clarify if the addition of another vector dilutes the expression of the prey protein. Densitometry measurements are made as before and analyzed statistically.

In summary, the search for mechanisms that regulate the many proteins in different cell systems can be tackled using a variety of different techniques. The results can provide you with a bounty of data that can be verified and used to mine many different databases. The outcomes from these experiments will provide you with new and fascinating insights that, heretofore, you may not have thought about. The critical issues are that you will need to obtain a clear understanding of what is occurring to the proteins at different stages of the experiment. This understanding will allow you to obtain clean representations of the proteins to be assessed with fewer inherent artifacts. Once mass spectrometry is completed, the data can then be used to mine various databases in order to fit and expand your data into an interactome.

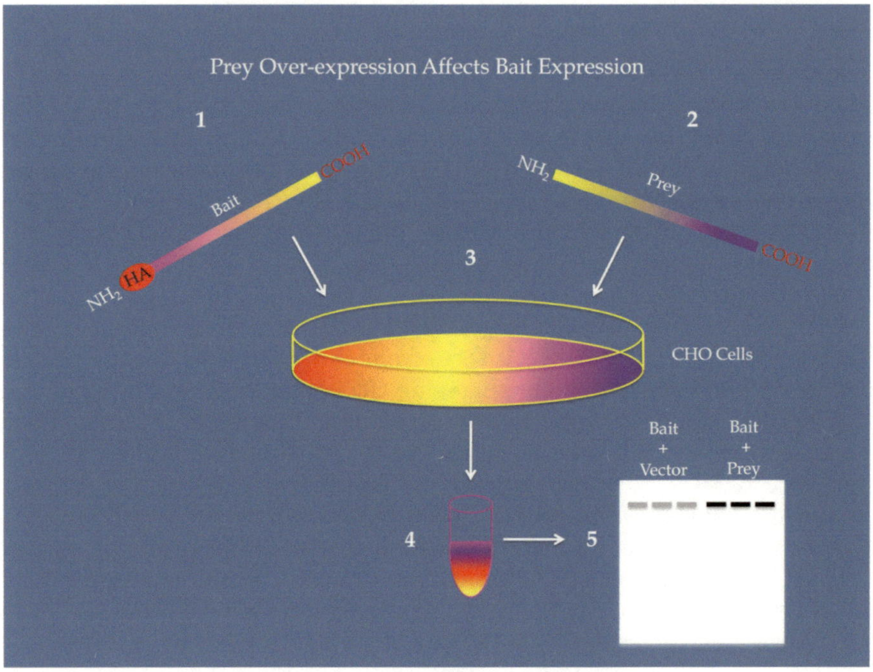

Fig. 4. Experiments, using cDNAs, are conducted to probe the function of protein-protein interactions when the prey is over-expressed in a heterologous expression system. **(1)** The original bait with the **(2)** newly-discovered prey or the bait with plasmid containing the reverse sequence of the prey (control) are cotransfected into **(3)** CHO cells. **(4)** Cells are collected and prepared for western blotting, where expression of the bait protein is probed **(5)** for both treatment and control conditions.

5. Bioinformatics

5.1 Comparison of high-throughput data with existing data

Once the initial experiments are completed, the lab worker will often wish to analyze the data by comparison to known interactions already existing in molecular interaction databases. These resources exist to collate and curate experimental data from laboratories around the world. Initially, interaction databases were established in isolation and often performed redundant curation of the same, high-visibility papers to their own standards, subsequently releasing the data in their own proprietary formats. However, more interaction data is published in any one calendar month than can be captured by these resources collectively, and no single database can claim to have complete coverage of the literature. In order to approach a complete interactome for the organism or process of interest, the user has always had to combine data from multiple resources. This was well nigh impossible prior to 2004, as the different data formats, required separate parsers to be written for each data source. This approach began to change with the release of the first HUPO-PSI standard representation of interaction data. Nowadays, two related formats exist - PSI-MI XML2.5 and MITAB2.5 (Kerrien et al., 2007), which are supported by both the majority of interaction databases and also by related visualization and analytical resources. These formats have enabled consistent data capture by multiple resources, with the choice of XML or tab-deliminated files often driven by either the complexity of data that the user wishes to harvest, or the amount of bioinformatic support available to them. Controlled vocabularies now make the terminology used to annotate these data consistent across the many data resources.

One major advantage of multiple resources sharing the same data format is that it is now possible to simultaneously access multiple resources with a single query, cluster the results, and visualize these in a single graph. A PSI Common Query InterfaCe (PSICQUIC) was developed that allows software clients to interact with multiple services and is based on the existing PSI MI file formats and the new Molecular Interaction Query Language (MIQL) (Aranda et al., in preparation). MIQL is based on standard Lucene syntax (http://lucene.apache.org/) and offers single word or phrase queries (abl1 AND "pull down"), search in specific data attributes/columns (abl1 AND species:human), wildcards (abl*), and logical operators. At the time of writing, 16 data providers were providing PSICQUIC servers, with a total of 16 million interactions available to query. PSICQUIC lays no constraint on data type or quality and much of what is available is also redundant, in that databases, which do not have their own curation team, will import information from those that do. To address this issue, several of the major databases have come together to synchronize their curation rules and data release through the IMEx Consortium (www.imexconsortium.org). This consortium allows the user to access and download a non-redundant, consistently annotated set of data, again using a PSICQUIC client to access appropriately tagged records (Orchard et al. in preparation).

5.2 Data resources

A number of databases exist, many of which have a bias in their curation strategy, either towards particular organisms or cellular processes. A brief summary of a number of these resources is given below, a more complete but less detailed list can be obtained from Pathguide (www.pathguide.org/). All the databases listed make their data available both through a dedicated website and also from their respective ftp sites in one or both of the PSI formats.

IntAct (www.ebi.ac.uk/intact) – no species or process bias, collects data from all organisms. Mainly contains protein-protein interactions but also annotates protein-small molecule, protein-nucleic acid. Interactions are derived from literature curation or direct user submissions and are freely available. Database and associated tools are open-source and available for download. IntAct provides a PSICQUIC service and is a full member of IMEx.

MINT (http://mint.bio.uniroma2.it/mint/) - no species or process bias, collects data from all organisms. Focuses on experimentally verified protein-protein interactions mined from the scientific literature. MINT provides a PSICQUIC service and is a full member of IMEx.

DIP (http://dip.doe-mbi.ucla.edu/dip) - no species or process bias, collects data from all organisms. Catalogs experimentally determined protein-protein interactions between proteins. The data are both manually curated and also automatically, using computational approaches that utilize knowledge about the protein-protein interaction networks extracted from the most reliable, core subset of the DIP data. DIP provides a PSICQUIC service and is a full member of IMEx.

MatrixDB (http://matrixdb.ibcp.fr) - focuses on interactions established by extracellular proteins and polysaccharides, mainly in human and mouse. Reports interactions of proteins, individual polypeptide chains or multimers (permanent complexes) and carbohydrates. MatrixDB provides a PSICQUIC service and is a full member of IMEx.

InnateDB (www.innatedb.ca/)- a database of the genes, proteins, experimentally-verified interactions and signaling pathways involved in the innate immune response of humans and mice to microbial infection. Integrates known interactions and pathways from major public databases together with manually-curated data. InnateDB provides a PSICQUIC service and is a full member of IMEx.

MPIDB (http://jcvi.org/mpidb/) - collects physical microbial interactions manually curated from the literature or imported from other databases. MPIDB provides a PSICQUIC service and is a full member of IMEx.

BioGrid (http://thebiogrid.org/) – a resource of protein–protein and genetic interactions for many model organism species. BioGRID provides a PSICQUIC service and is an observer member of IMEx.

HPRD (www.hprd.org/) – a database of human interactions, mixed species interactions (e.g. human-mouse) are modeled to human. Commercial entities have to pay a fee to use the data, under a licensing arrangement.

5.3 Data visualization

Once the user has downloaded their required data from one or all of the listed resources, the next step is to combine the networks then import and overlay your own data. The tool of choice for this exercise is most commonly Cytoscape (www.cytoscape.org), a free software package for visualizing, modeling and analyzing molecular and genetic interaction networks. In Cytoscape, nodes representing biological entities, such as proteins or genes, are connected with edges representing pairwise interactions, such as experimentally determined protein–protein interactions. Nodes and edges can have associated data attributes describing properties of the protein or interaction. Cytoscape allows users to extend its functionality by creating or downloading additional software modules known as 'plugins'. These plugins provide additional functionality in areas such as network data query, network data integration and filtering, attribute-directed network layout, Gene Ontology (GO) enrichment analysis and network motif functional module, protein complex or domain interaction detection.

Network data can be imported into Cytoscape is several formats, including those of the PSI-MI. To merge files in Cytoscape, the gene or protein identifier in the file must exactly match the corresponding Cytoscape node ID (or other Cytoscape attribute that has been previously loaded). If no matching identifiers are present, the situation can be corrected by loading an additional identifier into Cytoscape as a new node attribute. The ID mapping service supplied by UniProt (www.uniprot.org/) or by PICR (www.ebi.ac.uk/Tools/picr/) are recommended to achieve this correction. The Advanced Network Merge, a core Cytoscape plugin will allow network merging on the click of a single button.

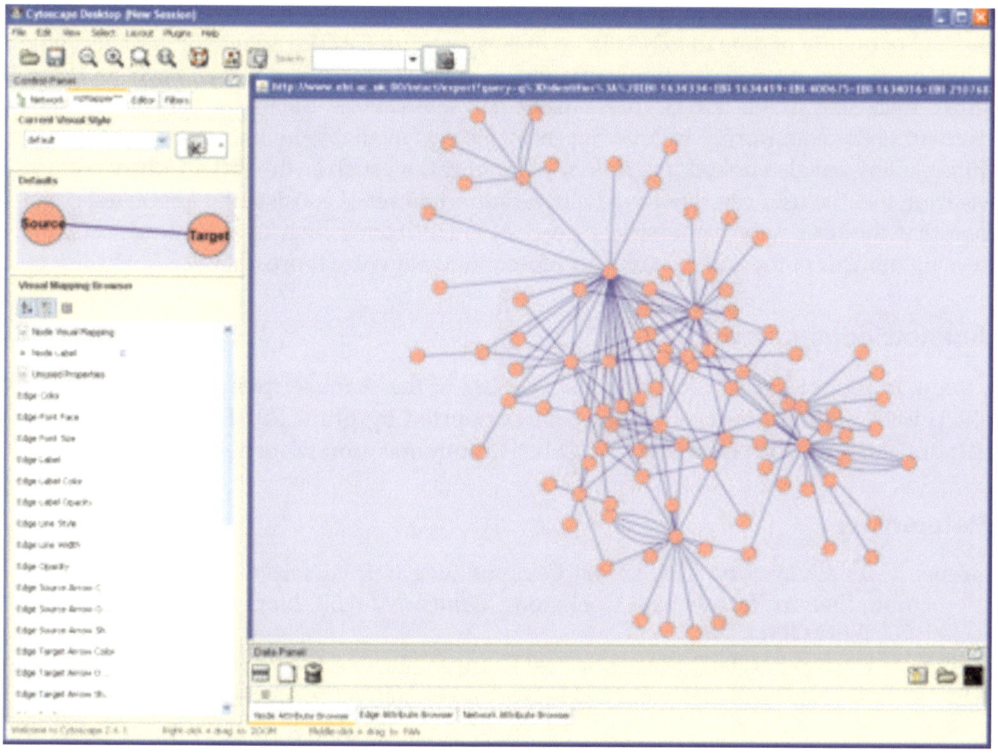

Fig. 5. An interaction network as viewed in Cytoscape.

5.4 Interolog mapping

Large-scale PPI networks are only available for a limited number of model organisms, therefore, groups working on less well-studied organisms have to rely on network inference using the interolog concept originally introduced by Walhout et al. (2000). This concept combines known PPIs from one or more source species and orthology relationships between the source and target species to predict PPIs in the target species. There are a number of resources available which perform the orthology mapping Inparanoid (http://inparanoid.sbc. su.se/cgi-bin/index.cgi) and Compara (www.ensembl.org /info/docs/api/ compara/index.html) being probably the best known. Few tools exist for interolog mapping, however, two database resources exist in which this exercise has been pre-computed for the user; STRING (http://string-db.org/) transfers associations/interactions between several hundred organisms and InteroPORC (http://biodev.extra.cea.fr/interoporc/) for the fully sequenced organisms described in the Integr8 database (www.ebi.ac.uk/integr8). Both resources make the data available in PSI format and both have a PSICQUIC server. Additionally, the InteroPORC software is freely available for in-house use (Michaut et al. 2008).

6. Conclusions

The data generated from experiments that examine genes and proteins has increased logarithmically over the last 20 years, largely driven by recent advances in high-throughput technologies that examine proteins individually, as well as in complexes. High-throughput protein studies that combine coimmunoprecipitations with 2-D gels have increased as a result of the higher quality of data obtained from mass spectrometry. The advent of these technologies has helped to fuel a need for the formation of many curated databases, such as those that capture molecular interactions. These molecular interactions databases are increasing their usefulness to the community by making their datasets available in a single, unified format. In addition, many are also linked in a unifying organization, such as the IMEx Consortium, which is ensuring that the user can download a non-redundant set of consistently annotated data. This means that the user now has a single point of entry from which to download data, and an increasing number of tools with which to subsequently analyze those data.

7. Acknowledgements

We thank Margaret Harvey for a critical reading of the manuscript and Sophia Sokolowski for help with the figures. This project was supported by grant R01-DC004295 to B.S. from the National Institute on Deafness and Other Communication Disorders.

8. References

Andrews, C.B., & Gregory, T.R. (2009) Genome Size is Inversely Correlated with Relative Brain Size in Parrots and Cockatoos. *Genome*, Vol.52, No.3 (March 2009), pp. 261–267, ISSN 0831-2796.

Ball, M.S., & Karuso P. (2007) Mass Spectral Compatibility of Four Proteomics Stains. *Journal of Proteome Research*, Vo.6, No.11 (November 2007), pp. 4313-4320, ISSN 1535-3893.

Bell, A.W., Deutsch, E.W., Au, C.E., Kearney, R.E., Beavis, R., Sechi, S., Nilsson, T., & Bergeron, J.J. & HUPO Test Sample Working Group (2009) A HUPO Test Sample Study Reveals Common Problems in Mass Spectrometry-Based Proteomics. *Nature Methods*, Vol.6, No.6, (June 2009), pp. 423-430, ISSN 1548-709.

Bradshaw RA. (2005) Revised Draft Guidelines for Proteomic Data Publication. *Molecular and Cellular Proteomics*, Vol.4, No.9, pp. 1223-1225, ISSN 1535-9476.

Chevallet, M., Diemer, H., Luche, S., van Dorsselaer, A., Rabilloud, T., & Leize-Wagner, E. (2006) Improved Mass Spectrometry Compatibility is Afforded by Ammoniacal Silver Staining. *Proteomics*, 2006, Vol.6, No.8 (April 2006), pp. 2350–2354, ISSN 1615-9853.

Cline, M.S., Smoot, M., Cerami, E., Kuchinsky, A., Landys, N., Workman, C., Christmas, R., Avila-Campilo, I., Creech, M., Gross, B., Hanspers, K., Isserlin, R., Kelley, R., Killcoyne, S., Lotia, S., Maere, S., Morris, J., Ono, K., Pavlovic, V., Pico, A.R., Vailaya, A., Wang, P.L., Adler, A., Conklin, B.R., Hood, L., Kuiper, M., Sander, C., Schmulevich, I., Schwikowski, B., Warner, G.J., Ideker, T., Bader, G.D. (2007), Integration of Biological Networks and Gene Expression Data Using Cytoscape. *Nature Protocols*, Vol. 2, No. 10 (September 2007), pp. 2366-2382, ISSN 1754-2189.

Duzhyy, D., Harvey, M., & Sokolowski, B. (2005) A Secretory-Type Protein, Containing a Pentraxin Domain, Interacts With an A-type K+ Channel. *Journal of Biological Chemistry*, Vol.280, No.15, (April 2005), pp. 15165-15172, ISSN 0021-9258.

Elbashir, S.M., Harborth, J., Weber, K., & Tuschl, T. (2002) Analysis of Gene Function in Somatic Mammalian Cells Using Small Interfering RNAs. *Methods*, Vol.26, No.2, (February 2002), pp. 199-213. ISSN 1046-2023.

Elbashir, S.M. Martinez, J., Patkaniowska, A., Lendeckel, W., & Tuschl, T. (2001) Functional Anatomy of siRNA for Mediating Efficient RNAi in *Drosophila melanogaster* Embryo

Lysate. *European Molecular Biology Organization Journal*, Vol.20, No.23, (December 2001), pp. 6877-6888, ISSN 0261-4189.

Eng, J.K., McCormack, A.L., & Yates III, J.R. (1994) An Approach to Correlate Tandem Mass Spectral Data of Peptides with Amino Acid Sequences in a Protein Database. *Journal of the American Society Mass Spectrometry*, Vol.5, pp. 976–989. ISSN 1044-0305.

Evan, G.I., Lewis, G.K., Ramsa, G., & Bishop, J.M. (1985) Isolation of Monoclonal Antibodies Specific for Human c-myc Proto-onco- Gene Product. *Molecular and Cellular Biology*, Vol.5 No.12, (December 1985), pp. 3610–3616. ISSN 0270-7306.

Fox, A.D., Hescott, B.J., Blumer, A., & Slonim, D.K. (2011) Connectedness of PPI Network Neighborhoods Identifies Regulatory Hub Proteins. *Bioinformatics*, Vol.27, No.8, (April 2011), pp. 1135-1142,, ISSN 1367-4803.

Gross J.H. (2004) Mass Spectrometry. Springer-Verlag, New York, NY 592 pp, ISBN 3540407391.

Golemis, E.A., & Adams, P.D. (2005) Protein-Protein Interactions, Cold Spring Harbor Laboratory Press, Cold Spring Harbor, NY, 938 pp, ISBN 0-87969-723-7.

Harvey, M.C., & Sokolowski, B.H.A. (2009) In Vivo Verification of Protein Interactions in the Inner Ear by Coimmunoprecipitation. In: *Auditory and Vestibular Research, Methods and Protocols*, B. Sokolowski (Ed.), Vol.493, No.II, 299-310, Humana Press, Springer Protocols. New York, NY, ISBN 978-1-934115-62-6.

Hopp, T.P., Prickett, K.S., Price V.L., Libby, R.T., March, C.J., Cerett, D.P., Urdal, D.L., & Conlon, P.J. (1988) A Short Polypeptide Marker Sequence Useful for Recombinant Protein Identification and Purification. *Nature Biotechnology*, Vol.6, (October 1988), pp. 1204–1210, ISSN 1087-0156.

Huber, L.A., Pfaller, K., & Vietor, I. (2003) Organelle Proteomics: Implications for Subcellular Fractionation in Proteomics. *Circulation Research*, Vol.92, No.9, (May 2003), pp. 962-968, ISSN 1524-4571.

Kathiresan, T., Harvey, M.C., & Sokolowski, B.H.A. (2009) The Use of 2-D Gels to Identify Novel Protein–Protein Interactions in the Cochlea. In: *Auditory and Vestibular Research, Methods and Protocols*, B. Sokolowski (Ed.), Vol.493, No.II, 269-286, Humana Press, Springer Protocols. New York, NY, ISBN 978-1-934115-62-6.

Kerrien S, Orchard S, Montecchi-Palazzi L, Aranda B, Quinn AF, Vinod N, Bader G, Xenarios I, Wojcik J, Sherman D, Tyers M, Salama JJ, Moore S, Ceol A, Chatr-aryamontri A, Oesterheld M, Stümpflen V, Salwinski L, Nerothin J, Cerami E, Cusick ME, Vidal M, Gilson M, Armstrong J, Woollard P, Hogue C, Eisenberg D, Cesareni G, Apweiler R, Hermjakob H (2007) Broadening the Horizon--Level 2.5 of the HUPO-PSI Format for Molecular Interactions. *BMC Bioinformatics* Vol.5, No.44 (October 2007), ISSN 1471-2105.

Lal, A., Haynes, S.R. & Gorospe, M. (2005) Clean Western Blot Signals From Immunoprecipitated Samples. *Molecular and Cellular Probes*, Vol.19, No.6, (December 2005), pp. 385–388, ISSN 0890-8508.

Mateo, C., Fernandez-Lorente, G., Pessela, B.C., Vian, A., Carrascosa, A.V., Garcia, J.L., Fernandez-Lafuente, R., & Guisan J.M. (2001) Affinity Chromatography of Polyhistidine Tagged Enzymes. New Dextran-Coated Immobilized Metal Ion Affinity Chromatography Matrices for Prevention of Undesired Multipoint Adsorptions. *Journal of Chromatography A*, Vol.915, No.1-2. (April 2001), pp. 97-106, ISSN 021-9673.

McGrath C.L. &. Katz, L.A. (2004) Genome Diversity in Microbial Eukaryotes. *Trends in Ecology and Evolution*, Vol.19, No.1 (January 2004), pp. 32-38, ISSN 0169-5347.

Michaut, M., Kerrien, S., Montecchi-Palazzi, L., Chauvat, F., Cassier-Chauvat, C., Aude, J.C., Legrain, P., Hermjakob, H. (2008) InteroPORC: Automated Inference of Highly

Conserved Protein Interaction Networks. *Bioinformatics*, Vol. 24, No.14 (July 2008), pp. 1625-1631, ISSN 1367-4803.

Moebius, J., Denker, K., & Sickmann, A. (2007) Ruthenium (II) Tris-Bathophenanthroline Disulfonate Is Well Suitable for Tris-Glycine PAGE But Not for Bis-Tris Gels. *Proteomics* Vol7, No.4, (February 2007), pp. 524–527, ISSN 1615-9853.

Moore, G.E. Cramming More Components Onto Integrated Circuits. *Electronics (Magazine)*, Vol.38, No.8, (April 1965), pp. 114-117, Available from ftp://download.intel.com /museum/Moores_Law/Articles-press_Releases/Gordon_Moore_1965_Article. pdf.

Navaratnam, D., (2009) Yeast Two-Hybrid Screening to Test for Protein-Protein Interactions inf the Auditory system. In: *Auditory and Vestibular Research, Methods and Protocols*, B. Sokolowski (Ed.), Vol.493, No.II, 299-310, Humana Press, Springer Protocols. New York, NY, ISBN 978-1-934115-62-6.

Noronha, S., Kaufman, J., & Shiloach, J. (1999) Streamline Chelating for Capture and Purification of Poly-His-Tagged Recombinant Proteins. *Bioseparation*, Vol.8, (1999), pp. 145–151, ISSN 0923-179X.

Olivier, M., Aggarwal, A., Allen, J., Almendras, A.A., Bajorek, E.S., Beasley, E.M., Brady, S.D., Bushard, J.M., & Stanford Human Genome Group (2001) A High-Resolution Radiation Hybrid map of the Human Genome Draft Sequence. *Science*, Vol.291, No.5507, (February 2001), pp. 1298-1302, ISSN 0036-8075.

Pedersen RA. (1971) DNA Content, Ribosomal Gene Multiplicity, and Cell Size in Fish. *Journal of Experimental Zoology*, Vol.177, No.1, (May 1971), pp. 65-78. ISSN 1932-5231.

Pellicer, J., Fay, M.F., & Leitch, I.J. (2010) The Largest Eukaryotic Genome of Them All? *Botanical Journal of the Linnean Society*, Vol.164, No.1 (September 2010), pp. 10–15, ISSN 0024-4074.

Richert, S., Luche, S., Chevallet, M., Van Dorsselaer, A., Leize-Wagner, E., & Rabilloud, T. (2004) About the Mechanism of Interference of Silver Staining with Peptide Mass Spectrometry. *Proteomics*, Vol.4, No.4 (April 2004), pp. 909–916, ISSN 1615-9853.

Sanger, F., Nicklen, S., & Coulson, A.R. (1977) DNA Sequencing with Chain-Terminating Inhibitors, *Proceedings of the National Academy of Sciences USA*, Vol.74, No.12, (December 1977), pp. 5463–5467, 0027-8424, ISSN 0027-8424.

Smejkal, G.B., Robinson, M.H., & Lazarev, A. (2004) Comparison of Fluorescent Stains: Relative Photostability and Differential Staining of Proteins in Two-Dimensional Gels. *Electrophoresis*, Vol.25, No.15, (August 2004), pp. 2511-2519, ISSN 0173-0835.

Terpe, K. (2003) Overview of Tag Protein Fusions: From Molecular and Biochemical Fundamentals to Commercial Systems. *Applied Microbiology and Biotechnology*, Vol.60, No.5, (January 2003), pp. 523-533, ISSN 0175-7598.

Venter, J.C., Adams, M.D., Myer, E.W., Li, P.W., Mural, R.J., Sutton, G.G., Smith, H.O., Yandell, M., & Celera Genome Working Group (2001) The Sequence of the Human Genome, *Science*, Vol.291, No.5507, (February 2001), pp. 1304-1351, ISSN 0036-8075.

Vetter, D.E., Basappa, J., & Turcan, S. (2009) Multiplexed Isobaric Tagging Protocols for Quantitative Mass Spectrometry Approaches to Auditory Research, In: *Auditory and Vestibular Research, Methods and Protocols*, B. Sokolowski (Ed.), Vol.493, No.II, 345-366, Humana Press, Springer Protocols. New York, NY, ISBN 978-1-934115-62-6.

Vinogradov, A.E. (2003) Selfish DNA is Maladaptive: Evidence From the Plant Red List. *Trends in Genetics*, Vol.19, No.11, (November 2003), pp. 609–614, ISSN 0168-9525.

Walhout, A.J., Boulton, S.J., Vidal, M. (2000) Yeast Two-Hybrid Systems and Protein Interaction Mapping Projects for Yeast and Worm. *Yeast*, Vol.17, No.2, (June 2000), pp.88-94, ISSN 0749-503X.

9

Flexible Protein-Protein Docking

Sebastian Schneider and Martin Zacharias
Technische Universität München/Physik Department T38,
85747 Garching,
Germany

1. Introduction

Biological processes almost always involve protein-protein interactions. Understanding the function of protein-protein interactions requires knowledge of the structure of the corresponding protein-protein complexes. The experimental structure determination by X-ray crystallography requires purification of large amounts of proteins. In addition, it is necessary to crystallize the proteins in the native complex which may not be feasible for all known interacting proteins. Multi-protein complexes mediate many cellular functions and are in a dynamic equilibrium with the isolated components or sub-complexes (Gavin et al., 2002; Rual et al., 2005). In particular, complexes of weakly or transiently interacting protein partners are often not stable enough to allow experimental structure determination at high (atomic) resolution. Experimental studies on detecting all protein-protein interactions in a cell indicate numerous possible interactions ranging from few to several hundred possible binding partners for one protein (Gavin et al., 2002; Rual et al., 2005). A full understanding of cellular functions requires structural knowledge of all these interactions. In the foreseeable future it will not be possible to determine the structure of all detected protein-protein interactions experimentally at high resolution. Structural modeling and structure prediction is therefore of increasing importance to obtain at least realistic structural models of complexes (Bonvin, 2006; Andrusier et al., 2008; Vajda & Kozakov, 2009; Zacharias, 2010). If the structure of the isolated protein partners or of closely related proteins is available it is possible to use a variety of computational docking methods to generate putative complex structures.

The driving force for the protein binding process corresponds to the associated change in free energy which depends on the structural and physicochemical properties of the protein partners. The "lock and key" concept of binding proposed by E. Fischer (Fischer, 1894) emphasizes the importance of optimal sterical complementarity at binding interfaces as a decisive factor to achieve high affinity and specificity. However, proteins and other interacting biomolecules are not rigid but can undergo a variety of motions even at physiological temperatures. The induced fit concept has evolved based on the observation that binding can result in significant conformational changes of partner molecules (Koshland, 1958). Within this concept protein partners induce conformational changes during the binding process that are required for specific complex formation. It should be emphasized, that in principle all possible molecular recognition processes require a certain degree of conformational adaptation. In recent years extensions of the induced-fit concept, based on ideas from statistical physics, emerged. A pre-existing ensemble of several inter-convertible conformational states being in equilibrium has been postulated (Tsai et al., 1999). Among these states are structures close to the bound and unbound forms. Binding of a partner molecule to the bound form shifts the equilibrium towards the bound form. Since every conformation is, in principle, accessible albeit with a potentially low statistical weight already in the unbound state the original induced fit concept is a special case of ensemble selection where only the presence of a ligand gives rise to an appreciable concentration of the bound partner structure.

Progress in protein-protein docking prediction methods has been monitored with the help of the community wide Critical Assessment of Predicted Interactions (CAPRI) experiment (Janin et al., 2003; Lensink et al., 2007). In this challenge participating groups test the performance of docking methods in blind predictions of protein-protein complex structures. The results of the CAPRI challenge indicate that for protein partners with minor conformational differences between unbound and bound conformation and some experimental hints on the interaction region often quite accurate predictions of complex structures are possible (Bonvin, 2006; Andrusier et al., 2008; Zacharias, 2010). However, the docking problem becomes much more difficult when protein partners undergo significant conformational changes upon association or for protein structures based on comparative modeling (Andrusier et al., 2008; Zacharias, 2010). The magnitude of possible conformational changes during association can range from local alterations of side chain conformations to global changes of domain geometries and can even involve refolding of protein segments upon association. Computational approaches to realistically predict protein-protein binding geometries need to account for such conformational changes. Often, protein-protein complex structures obtained from protein-protein docking but also in case of comparative modeling are of limited accuracy and require further structural refinement to achieve the generation of a realistic structural model. Since rigid docking is computationally much faster compared to flexible docking, the majority of current protein-protein docking approaches distinguishes between a first exhaustive systematic docking search followed by a second refinement step of pre-selected putative complexes (Bonvin, 2006; Vajda & Kozakov, 2009). Docking protocols may even consist of several consecutive refinement and rescoring steps (Andrusier et al., 2008). In the present contribution recent progress in the area of protein-protein docking with an emphasis on modeling conformational changes and adaptation during protein binding processes will be discussed.

2. Protein protein docking

The purpose of computational protein-protein docking methods is to predict the structure of a protein-protein complex based on the structure of the isolated protein partners. If the structure of the isolated partner proteins is not known it is often possible to build structures based on sequence homology to a known structure using comparative modelling methods.

Receptor and ligand proteins are discretized on three-dimensional grids and are portioned into inside, surface and outside regions, respectively. Matching of surfaces is measured by the overlap of surface regions. For each ligand rotation with respect to the receptor the correlation problem is solved using Fast-Fourier-Transformation (FFT). After filtering and possible refinement steps solutions with high overlap of surface regions (high surface complementarity) are collected as putative solutions.

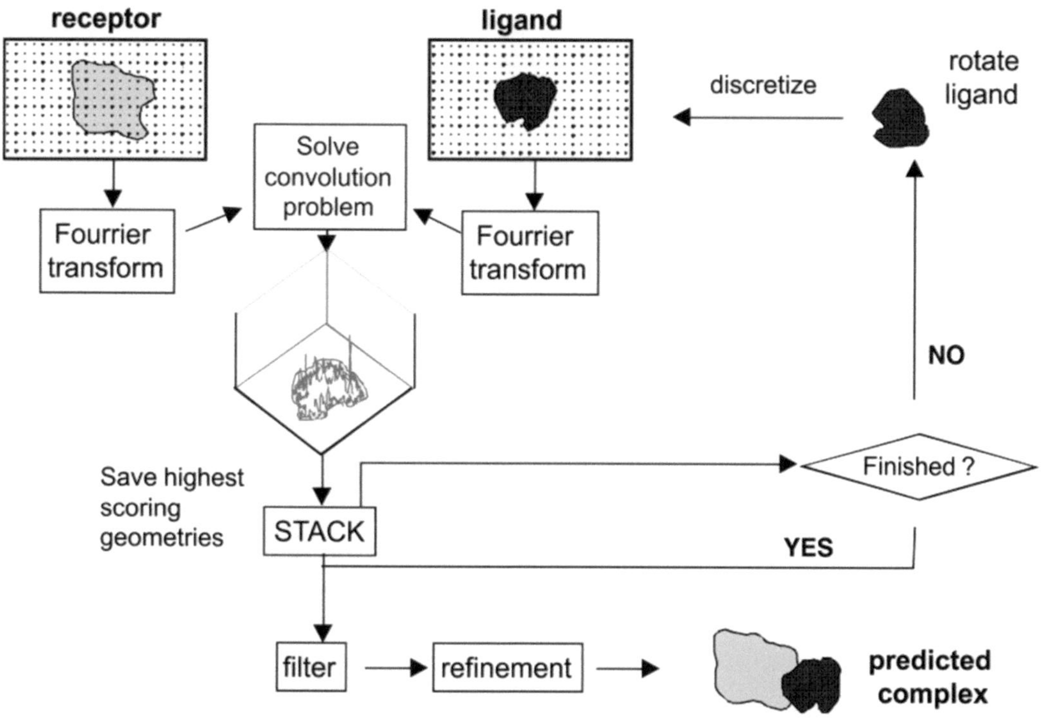

Fig. 1. Rigid protein-protein docking using Fast Fourier Transformation to solve a correlation problem.

2.1 Rigid docking methods

A variety of computational methods have been developed in recent years to efficiently generate a large number of putative binding geometries. The initial stage consists typically of a systematic docking search keeping partner structures rigid (Bonvin, 2006; Vajda & Kozakov, 2009). Subsequently, one or more refinement and scoring steps of a set of preselected rigid docking solutions are added to achieve closer agreement with the native geometry and to recognize near-native docking solutions preferentially as the best or among the best scoring complexes. In the initial search some unspecific sterical overlap between docking partners is typically tolerated to implicitly account for conformational adjustment of binding partners (e.g. Pons et al., 2009). Among the most common are geometric hashing methods to rapidly match geometric surface descriptors of proteins (Norel et al., 1994) and fast Fourier transform (FFT) correlation techniques to efficiently locate overlaps between complementary protein surfaces (Katchalski-Katzir et al., 1992). In the latter approach the two protein partners are represented by cubic grids, the grid points are assigned discrete values for inside, outside and on the surface of the protein. A geometric complementarity score can be calculated for the two binding partners by computing the correlation of the two grids representing each protein. Instead of summing up all the pair products of the grid entries one can make use of the Fourier correlation theorem. The corresponding correlation integral can easily be computed in Fourier space. The discrete Fourier transform for the receptor grid needs to be calculated only once. Due to the special shifting properties of Fourier transforms the different translations of the ligand grid with respect to the receptor grid can be computed by a simple multiplication in Fourier space. This process is repeated for various relative orientations of the two proteins. A disadvantage of standard Cartesian FFT-based correlation methods is the need to perform FFTs for each relative orientation of one protein molecule with respect to the partner. This can be avoided by correlating spherical polar basis functions that represent, for example, the surface shape of protein molecules. It has been successfully applied in the field of protein-protein docking (Ritchie et al., 2008). Recently, new multidimensional correlation methods have been developed that allow the correlation of multi-term potentials. Each function needs to be expressed in terms of spherical basis functions characterizing the surface properties of the protein partners (Ritchie et al., 2008; Zhang et al., 2009).

Geometric hashing is another common approach to identify possible protein-protein arrangements. It has been originally used as a computer visualization technique to match complementary substructures of one or several data sets (Norel et al., 1994). In protein-protein docking each protein surface is discretized as a set of triangles, which are stored in a hash table. By means of a hash key similar matching triangles on the surface of protein partners can be found quickly. During docking, these triangles comprise points on a molecular surface, having a certain geometrical (concave, convex) or physico-chemical (polar, hydrophobic) character. By matching triangles belonging to different molecules and being of complementary character, putative complex geometries can be generated.

A third class of methods uses either Brownian Dynamics (Schreiber et al., 2009; Gabdoulline & Wade, 2002), Monte Carlo, or multi-start docking minimization to generate large sets of putative protein-protein docking geometries (Zacharias, 2003; Fernandez-Recio et al., 2003; Gray et al., 2003). These methods have in principle the capacity to introduce conformational flexibility of binding partners already at the initial search step. Since these approaches are computational more expensive compared to FFT based correlation methods or geometric hashing a search is frequently limited to predefined regions of the binding partners (Bonvin, 2006). Alternatively, it is possible instead of atomistic models to employ coarse-grained (reduced) protein models to perform systematic docking searches. With such reduced protein models it is possible to optimize docking geometries starting from tens of thousands of protein start configurations (Zacharias, 2003; May & Zacharias, 2005). In order to limit the number of putative complex structures generated during an initial docking search cluster analysis is typically employed to reduce the number to a subset of representative complex geometries. Recently, the limitations of rigid docking strategies combined with a rescoring step have been systematically investigated by Pons et al. (2009). The authors applied a combination of rigid FFT-correlation based docking and re-scoring using the pyDock approach (Cheng et al., 2007). PyDock combines electrostatic Coulomb interactions with a surface-area-based solvation term (and an optional van der Waals term). The protocol showed very good performance for most proteins that undergo minor conformational changes upon complex formation (<1 Å Rmsd between unbound and bound structures) but unsatisfactory results for cases with significant binding induced conformational changes or applications that involved homology modelled proteins. A conclusion is that more specific scoring requires at the same time an improvement of the prediction accuracy of proposed binding modes in terms of deviation from the experimental binding interface. It also indicates the coupling between realistic scoring and accurate prediction of the complex structure.

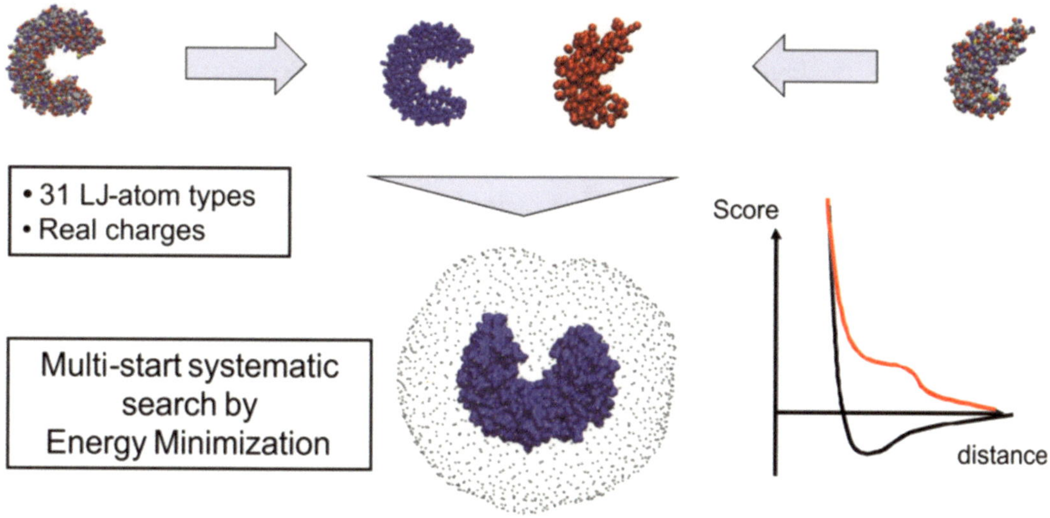

Fig. 2. Illustration of the ATTRACT docking methodology In the ATTRACT docking approach (Zacharias, 2003) atomic resolution partner structures are first translated (arrows) into a reduced (coarse-grained) representation based on pseudo atoms representing whole chemical groups. The smaller (ligand) protein is placed at various orientations on many starting placements around the receptor protein (in the middle of the Figure) followed by energy minimization to find an optimal docking geometry. In case of an attractive pseudo atom pair (black line in the plot on the right) an r^{-8}/r^{-6}-Lennard-Jones-type potential is used (r is the distance between atoms). For a repulsive pair (red curve) the energy minimum is replaced by a saddle point. The mathematical form of the scoring function is given in (Fiorucci & Zacharias, 2010b).

2.2 Flexible docking methods

A significant fraction of experimentally known protein-protein complexes belongs to the class that show only little conformational change upon complex formation. As indicated above in such cases it is possible to separate the initial rigid search from a subsequent flexible refinement and re-scoring step (see below). However, for many interesting docking cases with large associated conformational changes explicit consideration of conformational flexibility during the entire docking procedure or at an early refinement step appears to be necessary. Furthermore, in order to enhance the impact of docking in structural biology it is highly desirable to be able to use protein structures obtained by comparative (homology) modeling based on a known template structure with sufficient sequence similarity to the target protein. The accuracy of such comparative models depends on the correct alignment of target and template sequence. Even in cases of significant average target-template similarity the quality of the alignment is often not uniform along the whole protein sequence for example due to insertions or deletions in the aligned sequences which can result in structural inaccuracies. Overlap of such inaccurate structural segments with the protein region in contact with binding partners may interfere with the possibility to produce near-native complexes using rigid docking methods. This is also reflected in the fact that docking cases that involve homology modelled protein partners belong to the most difficult cases in the CAPRI docking challenge (Lensink et al., 2007).

One possibility to directly use computationally rapid rigid docking algorithms is to indirectly account for receptor flexibility by representing the receptor target as an ensemble of structures. The structural ensemble can, for example, be a set of structures obtained experimentally (e.g. from nuclear magnetic resonance (NMR) spectroscopy) or can be formed by several structural models of a protein. It is also possible to generate ensembles from MD simulations (Grunberg et al., 2004) or from distance geometry calculations (de Groot et al., 1997). Docking to an ensemble increases the computational demand and due to the large number of protein conformations may also increase the number of false positive docking solutions. In the field of small-molecule docking a variety of ensemble based approaches have been developed in recent years (reviewed in Totrov & Abagyan, 2008). Cross docking to ensembles from MD simulations have also been used to implicitly account for conformational flexibility in protein docking (Krol et al., 2007). Mustard & Ritchie (2005) generated protein structures deformed along directions compatible with a set of distance constraints reflecting large scale sterically allowed deformations. Subsequently, the structures were used in rigid body docking searches to identify putative complex structures. Conformer selection and induced fit mechanism of protein-protein association have been compared by ensemble docking methods using the RosettaDock approach (Chaudhury & Gray, 2008). The RossettaDock approach includes the possibility of modelling both side chain as well as backbone changes for a set of starting geometries obtained from a low-resolution initial search (Wang et al., 2007). The method was able to successfully select binding-competent conformers out of the ensemble based on favourable interaction energy with the binding partner (Chaudhury & Gray, 2008). It was recently shown that the Rosetta approach can also be used to simultaneously fold and dock the structure of symmetric homo-oligomeric complexes starting from completely extended (unfolded) structures of the partner proteins (Das et al., 2009).

For a limited number of start configurations (in case of knowledge of the binding sites) it is possible to combine docking with molecular dynamics (MD) or Monte Carlo (MC) simulations. This allows, in principle, for full atomic flexibility or flexibility restricted to relevant parts of the proteins during docking. The HADDOCK program employs MD simulations including ambiguous restraints to drive the partner structures towards the approximately known interface (Dominguez et al., 2003). The success of HADDOCK in many Capri rounds for targets where some knowledge of the interface region was available underscores also the benefits of treating flexibility explicitly during early stages of the docking process. For protein-protein docking it is always helpful to include some knowledge on the putative interaction region. In these cases the docking problem can often be reduced to the refinement of a limited set of docked complexes close to the known binding site. Fortunately, for proteins of biological interest and with experimentally determined structure there is often also some biochemical (e.g. mutagenesis) data available on residues involved in binding to other proteins. Alternatively, bioinformatics techniques to predict putative protein interaction regions can often be used to limit or restraint the docking search to relevant protein surface parts. Several new techniques to locate putative binding sites based on physico-chemical properties or evolutionary conservation have been developed in recent years (e.g. de Vries & Bonvin, 2008).

Protein partner structures can undergo not only local adjustments (e.g. conformational adaptation of side chains and backbone relaxation at the interface) during association but also more global conformational changes that involve for example large loop movements or domain opening-closing motions. Proteins in solution are dynamic and the question to what extend the accessible conformational space in the unbound form overlaps with the bound conformation has been at the focus of several experimental and computational studies. Elastic Network Model (ENM) calculations are based on simple distance dependent springs between protein atoms and despite its simplicity are very successful to describe the mobility of proteins around a stable state (Bahar et al., 1997; Bahar et al., 2006). Systematic applications to a variety of proteins indicate that there is often significant overlap between observed conformational changes and a few soft normal modes obtained from an ENM of the unbound form (Keskin, 1998; Tobi & Bahar, 2005; Bakan & Bahar, 2009). ENM-based normal mode analysis has been used to identify hinge regions in proteins (Emekli et al., 2008) and can also be used to design conformational ensembles.

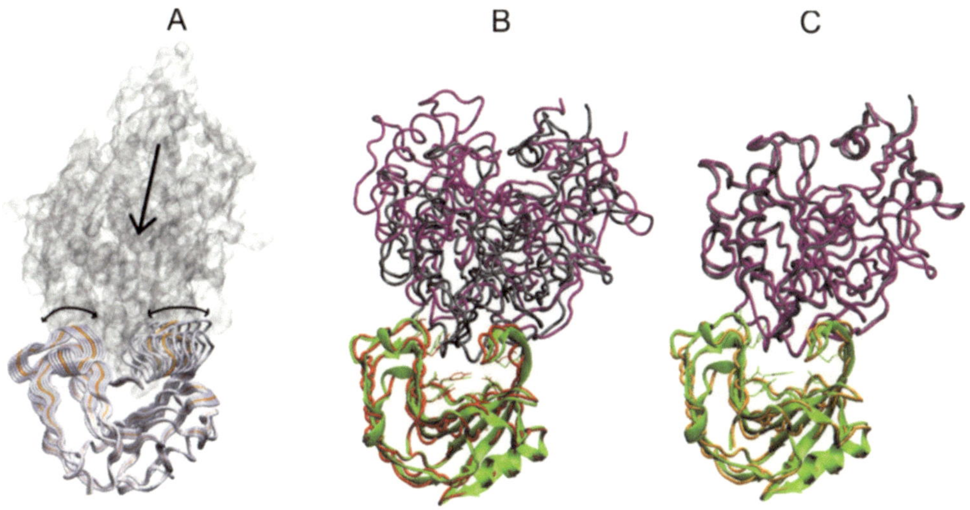

Fig. 3. Docking including minimization in soft flexible normal modes (A) Illustration of the flexible docking process of the taxi-inhibitor protein (pdb3HD8) to the xylanase target receptor protein (pdb1UKR) using the ATTRACT program (May & Zacharias, 2005). Putative translational motion of the inhibitor during docking approach is indicated by an arrow and the deformability of the xylanse by the superposition of several structures deformed in the softest normal mode (grey backbone tube representation). Best possible docking solutions (in pink) of the inhibitor relative to the bound (green cartoon) and unbound xylanase (red tube) are shown for rigid (B) and flexible (C) docking employing minimization along the 5 softest normal mode directions of the xylanase receptor protein. The placement of the inhibitor in the experimental structure is shown as grey tube. For flexible docking the root-mean-square deviation (Rmsd) from the inhibitor placement in the experimental structure was < 2 Å compared to > 8 Å in case of rigid docking.

It is also possible to use soft collective normal mode directions as additional variables during docking by energy minimization (Zacharias & Sklenar, 1999; May & Zacharias, 2005).

This allows the rapid relaxation of protein structures on a global scale involving much larger collective displacements of atoms during minimization then conventional energy minimization using Cartesian or other internal coordinates. The application of refinement in normal mode variables has been applied successfully in a number of studies (Lindahl & Delarue, 2005; May & Zacharias, 2005; Mashiach et al., 2010). Based on a coarse-grained protein model in the ATTRACT docking program (Zacharias, 2003) it has also been used in systematic docking searches to account approximately for global conformational changes already during the initial screen for putative binding geometries (May & Zacharias, 2008). In cases where protein partners undergo collective changes that overlap with the NM variables the approach can result in improved geometry and ranking of near-native docking solutions and can also lead to an enrichment of solutions close to the native complex structure (illustrated for an example case in Figure 3).

It should be emphasized that the inclusion of pre-calculated flexible degrees of freedom obtained from the unbound partners assumes that the collective directions of putative conformational change do not change upon binding to a partner protein. Although it has been shown that in many cases one can indeed describe a significant part of the observed conformational changes upon binding by a few collective degrees of freedom calculated for the unbound protein partners this does not need to be generally correct (Keskin, 1998; Tobi & Bahar, 2005; Bakan & Bahar, 2009). The binding partners may induce structural changes that are not possible for the isolated partner. In such cases pre-calculated flexible degrees of freedom cannot account for the true conformational change upon binding.

2.3 Prediction of putative binding regions prior to docking

If no experimental data on binding sites is available, binding site prediction methods can provide useful data for information driven docking. This type of information can be very helpful in order to limit the docking search or to evaluate and filter docking results. Docking approaches like HADDOCK (Dominguez et al., 2003) are based on applying restraints derived from experimentally known binding sites or predicted binding regions. Several different approaches exist to identify putative protein-protein binding sites. These methods focus on different characteristics of protein interaction sites like solvent accessibility (Chen & Zhou, 2005) or desolvation properties (Pons et al., 2009; Fiorucci & Zacharias, 2010a) and in many cases on combining different surface properties (Neuvirth et al., 2004; Liang et al., 2006). De Vries and Bonvin (2008) divided the properties of binding sites into three groups: 1. Properties of residues; 2. Evolutionary conservation; 3. Data obtained from atomic coordinates. The latter property includes, for example, secondary structure or solvent accessibility of residues or protein regions.

The data generated by predictors using one or more binding site features is presented either as a list of residues (Qin & Zhou, 2007) or as a patch on the proteins surface (Jones & Thornton, 1997a,b). Patch methods generate one or more patches of circular shape which can be found close to each other or distributed on the surface, sometimes additionally centre coordinates of these spots are given. In the other case residues from residue list predictors do not have to be nearby each other but are often clustered afterwards to receive a joined prediction at one or more spots on the proteins surface. Since proteins often have more than one binding site, prediction tools can indicate a correct binding site but maybe for the wrong binding partner.

Zhou & Qin (2007) and de Vries & Bonvin (2008) analysed existing predictors which are available as Web servers and evaluated the performance of these servers using 25 structures from the CAPRI targets and several other datasets. The binding site predictions can be used to evaluate possible predicted docking geometries but also to generate artificial binding sites around the prediction to bias the docking run towards a desired region. On the other hand predictions can be used to discard complexes with a low overlap of predicted contacts after a systematic docking run. Examples of predicted binding regions compared to the known binding sites are illustrated for two cases in Figure 4.

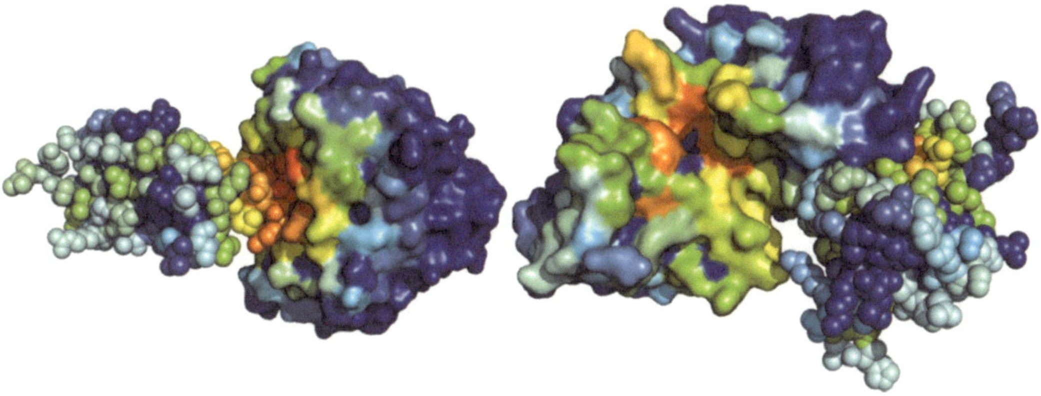

Fig. 4. Prediction of putative protein binding interfaces. Predictions were performed with the meta-PPISP server (Qin & Zhou, 2007) on the partner proteins of an enzyme inhibitor complex (pdb2SIC, left panel) and partners of a second complex (pdb1BUH, right panel). In each case one partner is represented as surface or collection of spheres, respectively. Protein partners are slightly displaced from the complexed state to indicate the native binding interface. Red indicates high predicted probability for a residue to be in the binding site and dark blue represents a low probability. Left example: The results match the real binding site. Right panel: The prediction for the smaller protein overlaps with the real binding site while for the larger protein residues quite far apart from the correct binding site are marked as putative binding site residues.

2.4 Flexible refinement and rescoring of docking solutions

As indicated in the two previous paragraphs protein-protein docking solutions obtained from an initial systematic docking run require typically a refinement and possibly also a rescoring step (Bonvin, 2006; Andrusier et al., 2008). This is not only necessary in case of rigid docking but also often if flexibility has been included approximately in the initial search by methods described in the previous paragraph (e.g. minimization in normal mode directions). The success of a multistep docking strategy requires that the set of initially docking structures contains solutions sufficiently close to the native structure in order to allow for further improvement during the refinement process. Hence, the initial scoring needs to recognize and preselect a binding mode sufficiently close the native placement and it has to simultaneously tolerate possible inaccuracies (atomic overlaps) at the interface. Before refinement the docking solutions are clustered to reduce the number of distinct docking geometries. Only one (the best scoring) solution from each cluster is typically used for further refinement and possible rescoring.

Refinement of a docked complex can be achieved by energy minimization based on a force field description of the proteins at atomic resolution. However, this results typically only in small displacements of atoms to minimize overlap and to optimize locally a hydrogen bonding network. Frequently, molecular dynamics (MD) simulations are employed to achieve larger conformational adjustments compared to energy minimization during docking refinement. MD simulations are based on numerically solving Newton's equation of motion in small time steps (1-2 fs = 1-2 10^{-15} s) based on a molecular mechanics force field description of the protein-protein complex (Karplus & McCammon, 2002). Due to the kinetic energy of every atom of the proteins, it is in many cases possible to overcome energy barriers and to move the structure significantly farer away from the initial docking geometry. Depending on the simulation temperature and length displacements up to several Angstroms from the initial atom positions are possible. However, if the displacements during MD simulations indeed move the proteins towards a more realistic complex structure depends on the accuracy of the force field and on a realistic representation of the aqueous solution. Refinement simulations on a given protein-protein complex should, ideally, include surrounding aqueous solvent and ions. This, however, increases the computational demand for such refinement simulations. In addition, the equilibration of explicit solvent molecules around a solute molecule requires significant simulation times (currently limited to tens or in some cases hundreds of nanoseconds). Nevertheless, during the final stages of some protein-protein docking protocols explicit water molecules can be added to the simulation system (van Dijk & Bonvin, 2006). Explicit solvent MD simulations can also be used to investigate the flexibility of protein structures prior to docking (Rajamani et al., 2004; Camacho, 2005). It is for example possible to identify the alternative or most likely side chain conformations. Using principal component analysis of the motions extracted from MD simulations it is also possible to analyse the global conformational flexibility of binding partners prior to docking (Amadei et al., 1993; Smith et al., 2005).

The possibility to implicitly account for solvent effects can be used to accelerate the refinement process. A variety of implicit solvation models has been developed (reviewed in Bashford & Case, 2000; Baker, 2005; Chen et al., 2008). Only a brief description of the most relevant concepts for protein-protein docking and scoring will be given. A macroscopic solvation concept describes the protein interior as a medium with a low dielectric permittivity embedded in a high dielectric continuum representing the aqueous solution (Baker, 2005). The effect of the solvent is then calculated as a reaction field from a solution of Poisson's equation for the charges assigned to each atom of the molecule. The mean effect of a salt atmosphere can be included by solving the Poisson-Boltzmann equation. The most common method to solve the Poisson-Boltzmann equation is the finite-difference method on a grid representation of the protein system. However, the method cannot easily be combined with MD refinement due to the difficulty to extract accurate solvation forces from grid solutions of the Poisson-Boltzmann equation (Gilson et al., 1993).

It is possible to use more approximate methods like the Generalized Born (GB) method (Still et al., 1990; Hawkins et al., 1995; Bashford & Case, 2000). In the GB approach an effective solvation radius is assigned to each atom. This effective radius can be thought of as an average distance of the selected atom from the solvent or from the solvent accessible surface of the molecule. With the effective Born radii calculated for each atom the electrostatic solvation and its derivative (solvation forces) can be calculated very rapidly (Schaefer & Karplus, 1996; Onufriev et al., 2002). The GB method and related implicit solvent approaches are frequently used during refinement of docked protein-protein complexes.

Once a set of docked and structurally refined complexes has been obtained a rescoring step can be used to finally select the most realistic predicted complex. An ideal scoring function

should recognize favourable native contacts as found in the bound complex and discriminate those from non-native contacts with lower scores. Scoring can be based on a physical force field with optimized weights on the energetic contributions (Dominguez et al., 2003; Bonvin, 2006; Audie, 2009) or can involve knowledge-based statistical potentials derived from known protein protein complex structures (Gottschalk et al., 2004; Zhang et al., 2005; Huang & Zou, 2008). Often a single descriptor (e.g. surface complementarity) or a single binding energy component (e.g. van der Waals or electrostatic energy) is non-optimal to distinguish non-native from near-native solutions. A combination of different surface and interface descriptors has been shown to better enrich near-native solutions in the pool of best scoring docking solutions (Murphy et al., 2003; Duan et al., 2005; Liu et al., 2006; Martin & Schomburg, 2008; Pierce & Weng, 2008; Audie, 2009;Liang et al., 2009).
The experimentally determined protein-protein complex structures allow the extraction of data on the statistics of residue-residue and atom-atom contact preferences at interfaces. Based on these statistics it is possible to design knowledge-based scoring functions which in general compare the frequency of contact pairs in known interfaces with the expected frequency if residues or atoms would randomly distributed at interfaces. Effective knowledge-based potentials have been developed that are based on contact preferences of amino acids at known interfaces compared to interfaces of non-native decoy complexes (Huang & Zou, 2008; Ravikant & Elber, 2009; Kowalsman & Eisenstein, 2009). The resulting contact or distance dependent pair-potentials can improve the scoring of near-native complexes. The distribution of amino acids in the core region of protein-protein interfaces differs on average from the whole interface and the rim region which is partially exposed to water even in the presence of the binding partner. This observation has also been explored to improve the recognition of near-native binding geometries and has been demonstrated on several test cases (Kowalsman & Eisenstein, 2009).

3. Conclusion

The rational modifications of protein surfaces are increasingly being used to design new protein-protein binding interfaces. Another ultimate aim of protein-protein docking approaches is the application on a systematic proteomic scale. Methods of protein-protein docking and interface refinement could help to predict possible protein interaction geometries and guide such protein interaction design. The realistic prediction of binding geometries of protein-protein complexes is highly desirable to provide structural models for the many important protein-protein interactions in a cell. Progress in both the efficiency and in the development of new docking algorithms has been achieved in recent years. Still a major challenge is the appropriate inclusion of possible conformational changes during the docking searches. This is of great importance since for the many protein interaction cases only homology modelled structures of the partners are available. Employing an appropriate ensemble of protein conformations or, alternatively, the efficient explicit consideration of conformational changes during docking are possible routes of progress. For many protein-protein interactions experimental data (e.g. low resolution structural or biochemical data) is available that restricts the range of possible complex structures. Here, restraint driven docking techniques that include flexibility of the binding partners at early refinement stages are promising. In recent years it has become clear that many protein-protein interactions involve coupled folding of disordered parts of proteins upon association. The possibility of structure prediction and modelling of such interactions is at a very early stage. Progress in this area will require many new algorithms and method developments.

4. Acknowledgment

This work has been support by the Deutsche Forschungsgemeinschaft (DFG grant Za153/5-3) and funding under the Sixth Research Framework Programme of the European Union (FP6 STREP "BacAbs", ref. LSHB-CT-2006-037325).

5. References

Amadei, A., Linssen, A. B. M. & Berendsen H. J. C. (1993). Essential dynamics of proteins, *Proteins* 17, 412-425.

Andrusier, N., Mashiach E., Nussinov R. & Wolfson H. J. (2008). Principles of flexible protein-protein docking, *Proteins: Structure, Function, and Bioinformatics* 73, 271-289.

Audie, J. (2009). Development and validation of an empirical free energy function for calculating protein-protein binding free energy surfaces, *Biophysical Chemistry* 139, 84-91.

Bahar, I., Atilgan, A.R., Erman, B. (1997). Direct evaluation of thermal fluctuations in proteins using a single-parameter harmonic potential, *Folding Design* 2, 173-181.

Bahar, I., Chennubhotla, C. & Tobi D. (2006). Intrinsic dynamics of enzymes in the unbound state and relation to allosteric regulation, *Current Opinion Structural Biology* 16, 194-200.

Bakan, A. & Bahar, I. (2009). The intrinsic dynamics of enzymes plays a dominant role in determining the structural changes induced upon inhibitor binding, *Proceeding of the National Academy of Sciences USA* 106, 14349-14354.

Bashford, D. & Case, D. A. (2000). Generalized born models of macromolecular solvation effects, *Annual Review of Physical Chemistry* 51 , 129-152.

Baker, N. A. (2005). Improving implicit solvent simulations: a Poisson-centric view, *Current Opinion Structural Biology* 15, 137-143.

Bonvin, A. M. J. J. (2006). Flexible protein-protein docking, *Current Opinion Structural Biology* 16, 194-200.

Camacho, C. J. (2005). Modeling side-chains using molecular dynamics improve recognition of binding region in CAPRI targets, *Proteins: Structure, Function, and Bioinformatics* 60, 245-251.

Chaudhury, S. & Gray, J. J. (2008). Conformer selection and induced fit in flexible backbone protein-protein docking using computational and NMR ensembles. *Journal of Molecular Biology* 381, 1068-1087.

Chen, J., Brooks, C. L. 3rd & Khandogin J. (2008). Recent advances in implicit solvent-based methods for biomolecular simulations, *Current Opinion Structural Biology* 18 , 140-148.

Chen, H. & Zhou, H.-X. (2005). Prediction of interface residues in protein–protein complexes by a consensus neural network method: Test against NMR data, *Proteins: Structure, Function, and Bioinformatics* 61, 21–35.

Cheng, T. M., Blundell, T. L. & Fernandez-Recio, J. (2007). PyDock: electrostatics and desolvation for effective scoring of rigid-body protein-protein docking, *Proteins* 68, 503-515.

Das, R., Andre, I., Shen, Y., Wu, Y., Lemak, A., Bansal, S., Arrowsmith, C.H., Szyperski, T., Baker, D. (2009). Simultaneous prediction of protein folding and docking at high resolution, *Proceeding of the National Academy of Sciences USA* 106, 18978-18983.

de Groot, B. L., van Aalten, D. M. F., Scheek, R. M., Amadei, A., Vriend, G., and Berendsen, H. J. C. (1997). Prediction of protein conformational freedom from distance constraints. *Proteins: Structure, Function, and Bioinformatics* 29, 240-251.

De Vries, S. J. & Bonvin, A. M. J. J. (2008). How proteins get in touch: interface prediction in the study of biomolecular complexes,. *Current Protein Peptide Science* 9, 394-406.

Dominguez, C., Boelens, R. & Bonvin, A. M. J. J. (2003). HADDOCK: A protein-protein docking approach based on biochemical and biophysical information, *Journal of the American Chemical Society* 125, 1731-1737.

Duan, Y., Reddy, B. V. B. & Kaznessis, Y. N. (2005). Physicochemical and residue conservation calculations to improve the ranking of protein-protein docking solutions, *Protein Science* 14, 316-328.

Emekli, U., Schneidman-Duhovny, D., Wolfson, H.J., Nussionov, R., Haliloglu, T. (2008). HingeProt: automated prediction of hinges in protein structures, *Proteins: Structure, Function, and Bioinformatics* 70, 1219-1227.

Fernandez-Recio, J., Totrov, M. & Abagyan, R. (2003). ICM-DISCO docking by global energy optimization with fully flexible side-chains, *Proteins: Structure, Function, and Bioinformatics* 52, 113-117.

Fiorucci, S. & Zacharias, M. (2010a). Prediction of Protein-Protein Interaction Sites Using Electrostatic Desolvation Profiles, *Biophysical Journal* 98, 1921-1930.

Fiorucci, S. & Zacharias, M. (2010b). Binding site prediction and improved scoring during flexible protein-protein docking with ATTRACT. *Proteins*. 78, 3131-3139.

Fischer, H. (1894). Einfluß der Configuration auf die Wirkung der Enzyme, *Chemische Berichte* 27, 2985-2993.

Gabdoulline, R. R. & Wade, R. C. (2002). Biomolecular diffusional association, *Current Opinion Structural Biology* 12, 204-213.

Gavin, A. C., Bosche, M., Krause, R., Grandi, P., Marzioch, M., Bauer, A., Schultz, J., Rick, J. M., Michon, A. M., Cruciat, C. M., et al. (2002). Functional organization of the yeast proteome by systematic analysis of protein complexes, *Nature*, 415, 141-147.

Gilson, M. K., Davis, M. E., Luty, B. A. & McCammon, J. A. (1993). Computation of Electrostatic Forces on Solvated Molecules Using the Poisson-Boltzmann Equation, *Journal of Physical Chemistry* 97, 3591-3600.

Gottschalk, K. E., Neuvirth, H. & Schreiber G. (2004). A novel method for scoring of docked protein complexes using predicted protein-protein interaction sites, *Protein Engneering* 17 183-189.

Gray, J. J., Moughon, S. E., Kortemme, T., Schueler-Furman, O., Misura, K. M. S., Morozov, A. V. & Baker, D. (2003). Protein-protein docking predictions or the CAPRI experiments, *Proteins: Structure, Function, and Bioinformatics* 52, 118-122.

Grunberg R., Leckner, J. & Nilges, M. (2004). Complementarity of structure ensembles in protein-protein docking, *Structure* 12, 2125-2136.

Hawkins, G. D., Cramer, C. J. & Truhlar, D. G. (1995). Pairwise solute descreening of solute charges from a dielectric medium, *Chemical Physics Letters* 246, 122-129.

Huang, S-Y., Zou, X. (2008). An iterative knowledge-based scoring function for protein-protein recognition, *Proteins* 71, 557-579.

Janin, J., Henrick, K., Moult, J., Eyck LT, Sternberg, M. J., Vajda, S., Vasker, I., Wodak, S. J. (2003). CAPRI: a critical assessment of predicted interactions, *Proteins : Structure, Function, and Bioinformatics* 52, 2-9.

Jones, S. & Thornton, J. M. (1997a). Analysis of protein-protein interaction sites using surface patches, *Journal of Molecular Biology* 272, 121-132.

Jones, S. & Thornton, J. M. (1997b). Prediction of protein-protein interaction sites using patch analysis, *Journal of Molecular Biology* 272, 133-143.

Karplus, M. & McCammon, J. A. (2002). Molecular dynamics of biomolecules. *Nature Structural Biolog,* 9 646-652.

Katchalski-Katzir, E., Shariv, I., Eisenstein, M., Friesem, A. A., Aflalo, C. & Vakser I. A. (1992). Molecular surface recognition: Determination of geometric fit between proteins and their ligands by correlation techniques, *Proceeding of the National Academy of Sciences USA* 89, 2195-2199.

Keskin, O. (1998). Binding induced conformational changes of proteins correlate with their intrinsic fluctuations: a case study of antibodies. *BMC Structural Biology* 7, 31-42.

Koshland, D. E. (1958). Application of a theory of enzyme specificity to protein synthesis, *Proceeding of the National Academy of Sciences USA* 17, 1145-1150.

Kowalsman, N. & Eisenstein M. (2009). Combining interface core and whole interface descriptors in postscan processing of protein-protein docking models, *Proteins: Structure, Function, and Bioinformatics* 77, 297-318.

Krol, M., Chaleil, R. A., Tournier, A. L. & Bates P. A. (2007). Implicit flexibility in protein docking : cross docking and local refinement, *Proteins: Structure, Function, and Bioinformatics* 69, 7-18.

Lensink, M. F., Mendez, R., Wodak, S. J. (2007). Docking and scoring protein complexes: CAPRI 3rd Edition, *Proteins Structure, Function, and Bioinformatics*, 69, 704-718.

Lindahl, E. & Delarue, M. (2005). Refinement of docked protein-ligand and protein-DNA structures using low-frequency normal mode amplitude optimization, *Nucleic Acids Research* 33, 4496-4506.

Liang, S., Zhang, C., Liu, S. & Zhou, Y. (2006). Protein binding site prediction using an empirical scoring function, *Nucleic Acid Research* 34, 3698-3707.

Liang, S., Meroueh , S. O., Wang, G., Qiu, C., Zhou, Y. (2009). Consensus scoring for enriching near-native structures from protein-protein docking decoys, *Proteins: Structure, Function, and Bioinformatics* 75, 397-403.

Liu, S., Li, Q. & Lai, L. (2006). A combinatorial score to distinguish biological and non-biological protein-protein interfaces, *Proteins: Structure, Function, and Bioinformatics* 64, 68-78.

Martin, O., Schomburg, D. (2008). Efficient comprehensive scoring of docked protein complexes using probabilistic support vector machines. *Proteins: Structure, Function, and Bioinformatics* 70, 1367-1378.

Mashiach, E., Nussinov, R. & Wolfson. H. J. (2010). FiberDock: Flexible induced-fit backbone refinement in molecular docking, *Proteins: Structure, Function, and Bioinformatics* 78, 1503-1519.

May, A. & Zacharias, M. (2005). Accounting for global protein deformability during protein–protein and protein–ligand docking, *Biochimica et Biophysica Acta* 1754, 225-231.

May, A. & Zacharias, M. (2008). Energy minimization in low-frequency normal modes to efficiently allow for global flexibility during systematic protein-protein docking, *Proteins* 70, 794-809.

Murphy, J., Gatchell, D. W., Prasad, J. C. & Vajda S. (2003). Combination of scoring functions improves discrimination in protein-protein docking, *Proteins* 53, 840-854.

Mustard, D. & Ritichie, D. W. (2005). Docking essential dynamics eigenstructures, *Proteins: Structure, Function, and Bioinformatics* 60, 269-274.

Neuvirth, H., Raz, R. Schreiber, G. (2004). ProMate: A Structure Based Prediction Program to Identify the Location of Protein–Protein Binding Sites, *Journal of Molecular Biology* 338, 181-199.

Norel, R., Fischer, D., Wolfson, H. J. & Nussinov R. (1994). Molecular surface recognition by a computer vision-based technique, *Protein Engneering*. 7, 39-46.

Onufriev, A., Case, D. A. & Bashford, D. (2002). Effective Born radii in the generalized Born approximation: the importance of being perfect, *Journal of Computational Chemistry* 23, 1297-1304.

Pierce, B., Weng, Z. (2008). A combination of rescoring and refinement significantly improves protein docking performance, *Proteins: Structure, Function, and Bioinformatics* 72, 270-279.

Pons, C., Grosdidier, S., Solernou, A., Perez-Cano, L. & Fernandez-Recio, J. (2009) Present and future challenges and limitations in protein-protein docking, *Proteins: Structure, Function, and Bioinformatics*, 78, 95-108.

Qin, S. & Zhou, H.-X. (2007). meta-PPISP: a meta web server for protein-protein interaction site prediction, *Bioinformatics 23, 3386-3387.*

Rajamani, D., Thiel, S., Vajda, S. & Camacho, C. J. (2004). Anchor residues in protein- protein interactions, *Proceeding of the National Academy of Sciences USA* 101, 11287-11292.

Ravikant, D. V. S, Elber, R, (2009). PIE-efficient filters and coarse grained potentials for unbound protein-protein docking. *Proteins: Structure, Function, and Bioinformatics*, 75, 133-145.

Ritchie, D., Kozakov, D., Vadja, S. (2008). Accelerating and focusing protein-protein docking correlations using Multi-dimensional rotational FFT generating functions. *Bioinformatics* 24, 1865-1873.

Rual, J. F., Venkatesan, K., Hao, T., Hirozane-Kishikawa, T., Dricot, A., et al. (2005). Towards a proteome-scale map of the human protein-protein interaction network, *Nature*, 437, 1173-1178.

Schaefer, M. & Karplus, M. (1996). A comprehensive analytical treatment of continuum electrostatics, *Journal of Physical Chemistry* 100, 1578-1599.

Schreiber, G., Haran, G., Zhou, H.-X (2009) Fundamental Aspects of protein-protein association kinetics, *Chemical Reviews* 1009, 839-860.

Smith, G. R., Sternberg, M. J. E. & Bates P. A. (2005). The relationship between the flexibility of proteins and their conformational states on forming protein-protein complexes with an application to protein-protein docking. *Journal of Molecular Biology* 347, 1077-1101.

Still, W. C., Tempczyk, A., Hawley, R. C. & Hendrikson, T. (1990). Semianalytical Treatment of Solvation for Molecular Mechanics and Dynamics, *Journal of the American Chemical Society* 112, 6127-6129.

Totrov M. & Abagyan, R. (2008). Flexible ligand docking to multiple receptor conformations: a practical alternative, *Current Opinion Structural Biology* 18:178-184.

Tsai, C. J., Kumar, S., Ma, B. (1999). Nussinov R: Folding funnels, binding funnels, and protein function, *Protein Science* 8, 1181-1190.

Tobi, D., & Bahar, I. (2005). Structural changes involved in protein binding correlate with intrinsic motions of proteins in the unbound state, *Proceeding of the National Academy of Sciences USA* 2005, 102:18908-18913.

van Dijk, A. D. J. & Bonvin, A. M. J. J. (2006). Solvated docking: introducing water into the modelling of biomolecular complexes. *Bioinformatics*. 22, 2340-2347.

Vajda, S. & Kozakov D, (2009) Convergence and combination of methods in protein-protein docking. *Current Opinion Structural Biology* 19:164-170.

Wang, C., Bradley, P. & Baker, D. (2007). Protein-protein docking with backbone flexibility, *Journal of Molecular Biology* 373, 503-519.

Zacharias, M. & Sklenar, H. (1999). Harmonic modes as variables to approximately account for receptor flexibility in ligand-receptor docking simulations: Application to a DNA minor groove ligand complex, *Journal of Comput Chemistry* 20, 287-300.

Zacharias, M. (2003). Protein-protein docking using a reduced model, *Protein Science* 12, 1271-1282.

Zacharias, M. (2010). Accounting for conformational changes during protein-protein docking,. *Current Opinion Structural Biology* 16, 194-200.

Zhang, C., Liu, S. & Zhou, Y. (2005). Docking prediction using biological information, ZDOCK sampling technique, and clustering guided by the DFIRE statistical energy function,. *Proteins: Structure, Function, and Bioinformatics* 60, 314-318.

Zhang, Q., Sanner, M. & Olson A. J. (2009). Shape complementarity of protein-protein complexes at multiple resolutions, *Proteins: Structure, Function, and Bioinformatics* 75, 453-467.

Zhou, H.-X. & Qin, S. (2007). Interaction-site prediction for protein complexes: a critical assessment, *Bioinformatics* 23, 2203-2209.